F. M. L. Gillelen

The Oil Regions of Pennsylvania

with maps and charts of Oil Creek, Allegheny River, etc - Vol. 1

F. M. L. Gillelen

The Oil Regions of Pennsylvania
with maps and charts of Oil Creek, Allegheny River, etc - Vol. 1

ISBN/EAN: 9783337302658

Printed in Europe, USA, Canada, Australia, Japan

Cover: Foto ©Andreas Hilbeck / pixelio.de

More available books at www.hansebooks.com

THE

OIL REGIONS

OF

PENNSYLVANIA:

WITH

MAPS AND CHARTS OF

OIL CREEK, ALLEGHENY RIVER, ETC.

A Thoroughly Reliable Work, carefully Compiled.

By F. M. L. GILLELEN.

PITTSBURG, PENN.:
JOHN P. HUNT, PUBLISHER,
59 FIFTH STREET, MASONIC HALL.
LONDON: TRIEBNER & CO., 60 PATERNOSTER ROW.

Entered according to Act of Congress, in the year 1864, by
JOHN P. HUNT,
In the Clerk's Office of the District Court of the United States for the Western District of Pennsylvania.

STEREOTYPED AT THE
FRANKLIN TYPE FOUNDRY,
CINCINNATI, O.

PREFACE.

WHEN a new discovery has been announced to the world, we always have a share of tolerably educated people whose names and conversational style frequently introduce them into society. This class are known by their wholesale negativing of all things new. They have not succeeded this time. The universal household use of oil, with its admirable properties, compared to other fulminating liquids, has made it a staple now necessary to every family.

The discovery and development of petroleum, as described in the text of our work, was made at an early date by the aboriginals of the country, who used it for civic and religious purposes.

At a later period, and yet in the recollection of many middle-aged men, the oil was gathered and published as having rare medicinal qualities. The fifty-cent bottles carried large labels, advertising petroleum as a cure-all for "all the ills which mankind might be heir to."

Its sale, however, as a cure-all, was soon eclipsed by the industry and capital which developed it, and threw it into market for light and lubricating purposes; and which now forms a wealthy article of export, commensurate with other materials of America.

We publish this work in order to delineate the real oil region of Pennsylvania. The importance and wealth of this region, contributing to the necessaries of every family, is as yet but poorly comprehended.

We furnish a very accurate map of Oil Creek, together

with its tributaries. The Allegheny River, carefully delineated, with towns, villages, manufacturing establishments of iron and salt, and oil refineries, are especially noticed. This information is an especial desideratum to a stranger.

The work abounds in illustrations, while the statistical tables furnish an accurate list of oil companies, their stock, etc., together with exportations from Eastern cities.

To have an accurate account of the oil regions, with maps strictly drawn, has required labor, and it is hoped that a generous public will appreciate the efforts of the editor and publisher to distribute information concerning a necessary household comfort.

CONTENTS.

	PAGE
Allegheny River	7
Big Brokenstraw Creek	26
Big Brokenstraw Island	27
Big Scrubgrass Island	40
Black Fox Island	46
Bald Eagle Island	46
Bull Creek Island	55
Cherry Run	13
Cherry Tree Run	18
Clarke's Island	28
Courson Islands	30
Cogsley's Island	52
Crooked Creek Islands	52
Dale's Island	33
Evault's Defeat Island	42
Emlenton	44
Early's Island	49
Franklin	21
Freeport	23
Fourteen-Mile Island	56
Goose Flat Island	30
Hemlock Islands	32
Hickory Town Island	33
Holman's Island	34
Hemlock Creek Islands	35
Horse Creek Island	37
Hare's Island	57
Jackson's Island	26
J. Thompson's Island	27
Jack's Island	55
Kittanning	22
Karn's Island	55
Mead's Island	25
Mill stone Island	30

(v)

CONTENTS.

	PAGE
Maple Island	35
McCray's Island	36
McDowell's Island	38
Mahoning Islands	50
Murphy's Island	54
Mad-Dog Island	54
Nicholson's Islands	53
Nine-Mile Island	56
New York Oil Companies	65
Oil Regions	11
Oil City	14
Oil Creek	16
Oil Creek Island, No. 1	37
Oil Creek Island, No. 2	38
Plumerville	13
Pithole Island	36
Puckerty Island	56
PITTSBURG—Its Early History	58
Its Water and Railroad Facilities	59
Population and Business	60
Business Character	61
Petroleum	62
Pittsburg Oil Companies	64
Philadelphia Oil Companies	65
Robert Thompson's Island	28
Sharpsburg	24
Scott's Island	26
Steward's Islands	29
Shafer's Island	38
Steen's Island	39
Stover's Island	43
Stump Creek Islands	45
Six-Mile Island	57
Tidioute	19
Tarentum	24
Tidioute Island	31
Tionesta Islands	34
Two-Mile Run Island	38
Tar's Island	51
White Oak Island	31
Walnut Islands	36

ALLEGHENY RIVER.

The Allegheny River rises in the northern part of Pennsylvania, passes through the edge of New York, then winds its way back into Pennsylvania. It meanders through Warren County to the extent of fifty miles, and, when in Venango County, it directs its course to every point of the compass. Probably no river in the world rolls for the same distance so strange a current. Its name, Allegheny, was given by the Seneca tribe of Indians. Its etymology gives the definition "Fair Water." The French, during their occupation of the country, hailed its flowing current as "*La Belle Riviere.*" The wild and rugged country along its banks, the high and precipitous hillsides, often rising into bluffs and cliffs, present the grand and picturesque of nature's parentage. As it flows along, now southward, again almost directly north, it forms curves which add grace and majesty to itself. It courses its way through the several counties of Warren, Venango, Clarion, Armstrong, and Allegheny, whose agricultural, coal, and oil products find their way to market by the navigation which this useful river affords. It has many tributaries, small, and apparently insignificant; yet along its banks enterprise and capital have developed wealth—wealth in lumber, coal, iron, and oil. Small rivulets, tracing their way through mountain defiles, have been surveyed, the banks and meadows prospected, and, erelong, the smoke of the furnace or the thundering of the forge-hammer will add not only to the scenery, but to the wealth of the nation. The ax of the lumberman re-echoes through the lofty pines; the saw-mill, with its measured click, hightens the bewilderment of a visitor who may ascend a rugged cliff to survey the grandeur of the surrounding regions. While sixty to eighty million feet of lumber and eighty millions of shingles descend the river, it is suggested, "Are not the supplies exhausted?" Visit the regions; see the supply. "Inexhaustible!" you respond. The ax and its ally, the saw-mill, are as industrious as when the first raft slowly wound its way to the Ohio, to exchange its unhewed logs for the common staples of life. But while wealth accumulated slowly, and prosperity gradually opened its portals, through strong, industrial effort, in iron and lumber, a new source of wealth was auspiciously opened—a product for the world's consumption—an article of positive household necessity—the petroleum of the Alleghany regions. To-day its necessity for light and lubricating purposes are not only recognized at home, but the most distant foreign countries enlarge their orders. Thus the rugged districts through which the Alleghany and its tributaries run are made important aids toward the progress of the age.

We have, in our work, attempted to define the boundary line of present discoveries, and have given a chart of the famous Oil Creek, the contributions of which are now amazing the most latent of capitalists. We have likewise given a chart of the Alleghany River, with its tributaries, its most important towns

and villages, names of large iron furnaces, salt-works, and other places whose magnitude of business invites attention, as a guide to those who are strangers to this wealthy district. Oil Creek is particularly defined in the chart, but it may not be amiss to give a short historical account of it. The discovery of oil can not well be claimed by the present generation. Upon investigation, we find that the creek was named by the Seneca Indians, on account of a peculiar kind of inflammable oil found floating upon the surface of the water. They used it as an unguent, and, tradition says, often in their religious festivals. It was as celebrated with the Indian nations as the naphtha of the Caspian Sea. With it they mixed their war-paint, which gave them a hideous, glistening appearance, and added great permanency to the paint, as the oil rendered it impervious to water.

A commandant of Fort Duquesne writes to General Montcalm, the unfortunate hero of Quebec, in 1753, thus:

"I would desire to assure your Excellency this is a most delightful land. Some of the most astonishing natural wonders have been discovered by our people. While descending the Allegheny fifteen leagues below the mouth of the Conewango, and then above Fort Venango, we were invited, by the chief of the Senecas, to attend a religious ceremony of his tribe. We landed and drew up our canoes on a point where a small stream entered the river. The tribe appeared unusually solemn. We marched up the stream about half a league, where the company had arrived some days before us. Gigantic hills begirt us on every side. The scene was sublime. The surface of the stream was covered with a thick scum, which burst into a complete conflagration. The oil had been gathered and lighted by a torch. The Indians gave forth a triumphant shout, that made the hills and valleys re-echo again and again."

Centuries ago acquaintance was held with this material; but it remained for the present circumspective, investigating age to develop it, and introduce it into the households of one-half of the civilized nations.

How important, then, is the Allegheny, with its tributaries, upon the waters of which may be seen hourly boat after boat, covered with the iron-hooped barrels, calmly descending its waters, to contribute this great necessary of domestic and mechanical life!

We need not here describe the towns situated along the banks of this useful stream. In the text we have given the characteristics of each. Suffice it to say, How could they be otherwise than flourishing?

The time is yet in the recollection of many when this portion of Pennsylvania was noted for its sterility, compared to the rich agricultural valleys of Eastern Pennsylvania. To-day no more wealthy region can be found in the United States. The mineral and oil products are sources of fabulous wealth, and enterprise and activity developing them to their maximum yield.

For the cheap and facile transportation of all these immense products, let us give the Allegheny the credit to which she is entitled, and yet not forget her virgin name, "La Belle Riviere."

THE OIL REGIONS.

The oil, as it comes from the earth, either naturally, or is expelled by artificial means, forms one of the discoveries of the age, which tends so much to the comforts of mankind. Much theorizing has been done in regard to its geological parentage. While the territory over which it manifests itself has been visited and explored by amateur as well as professional geologists, an acknowledged difference of theory, as to its fountain sources, producing new theories, until we have them multiplied to an unusual extent. One thing we know, oil abounds in this portion of Pennsylvania, and capital is making it yield to a profitable extent. Quantities sufficient to supply the nation with what is now an indispensable requisite of every household furnished by the enterprise of individuals and companies.

The character of the soil, with oil localities, may be noticed without committing ourselves to any particular theory about its origin, which would be foreign to a work of this kind.

Geological surveys have discovered petroleum in the anthracite and calciferous beds of New York, and traces of it even in quartz-crystal regions. The limestone of Canada exudes small quantities, while the fossil coral at Watertown, New York, gives evidence of oil. Again, it occurs on the surface of a spring, and issues from the Utica slate, on Great Manitouline Island, on Lake Huron. At Black Rock, in the Niagara River, we have the corniferous beds, where it occupies cavities in fossiliferous formations, and in sufficient abundance to be an object of commerce. It is said that in Enniskillen, in Western Canada, there is a deposit of solid bitumen, or mineral tar, half an acre in extent, below which are limestone beds. In Erie, Seneca, and Cattaraugus Counties, New York, the rocks of the Chemung period (by which geologists convey a sandstone formation) afford abundant oil-springs. At Chicago, Illinois, recent explorations have discovered traces of oil. Parties are now engaged in further and more extensive efforts to develop the territory.

The oil-wells of the Allegheny and its tributaries are found in subcarboniferous sandstone, often descending through overlying carboniferous strata. This region of Pennsylvania seems to be the only place from which large quantities have been realized. In all other attempts small quantities have been obtained; but here can be found wells which flow from fifty to eight hundred barrels daily, while the region of Oil Creek itself exports eight thousand barrels daily. There can be no doubt but this region is the basin of petroleum. The outer edges may yield it in quantities to excite speculative movements, but the recompense in oil appears to be small, compared to that of Venango, Green, and Fayette Counties, of Pennsylvania.

Petroleum is a compound of hydrogen and carbon. Chemists vary its composition between $C_{18} H_{20}$ and $C_{26} H_{23}$. In becoming inspissated it is often more or less oxydized, losing sometimes, in part, its fusibility and its solubility in ether.

Within the past year it is estimated that no less than two hundred oil companies, representing an aggregate of nearly one hundred millions of dollars, have been organized, principally for operation in the Pennsylvania oil regions. Many of these companies are now successfully operating flowing wells in the Venango district.

Cherry Run

Exhibits marked evidences of good wells. There has not yet been a failure to obtain oil on Cherry Run. Many companies, with capitals ranging from $150,000 to $500,000, are busily at work sinking wells along this territory, some of which have secured remunerative wells. The greatest flowing well is about half a mile from the mouth of the river. Its yield is about two hundred and fifty barrels daily.

The contiguous territory gives evidence of oil, and much of it is now ready for lease. This run seems to be immediately upon the great basin of oil lying beneath the territory of about fifty square miles, and is esteemed, by men who have had much experience in oil-well explorations, to be profitable to those who sink wells.

The course of the run is almost parallel with Oil Creek. In the village of Plumerville and its vicinity, the sudden erection of derricks, visible every-where, gives evidence of the faith of companies and capitalists in expending time and money on its territory.

There is, consequently, a great excitement in this region. The discovery of a two-hundred-and-fifty-barrel well per diem draws adventurers; and the lands contiguous, as well as those more remote, have been quickly leased, and derricks every-where rearing their tall heads, while speculators in land and oil may be seen traversing the roads, eager to engage which may seem profitable, in their judgment.

Oil City.

Oil City is situated on the Allegheny River, at the mouth of Oil Creek, one hundred and fifty miles above Pittsburg. Oil Creek divides the place, over which there is a bridge connecting the two parts of the city.

It is the port for the oil of the territories for some miles above on the creek, and forms the chief harbor for oil shipments from the Oil Creek, Cherry Run, and Cherry-tree Run territories.

The immense quantities of oil upon the landings at certain seasons, awaiting shipment, present, in themselves, a curiosity for strangers, while wagon after wagon—estimated to be from one thousand to fifteen hundred daily—arrive and traverse its streets, laden with the products of the upper territory, astonish the stranger. For its commanding position in the oil trade it deservedly commands its name, and is entitled to the attention of visitors as well as capitalists.

In its main streets flowing wells, as well as pumping wells of profitable yield, are now in operation, while here and there new explorations are daily making. The different landings from which shipments are made, with their wonderful heaps of barrels, and fleets of boats lading and unlading, present a scene of busy activity rarely witnessed in a place of similar size.

The resident population is computed to be about four thousand, and steadily increasing; and it is not unreasonable to believe that, as further developments are made, there will continue to be drawn hither capitalists from Europe as well as our own country, ready to invest in this city of the oil trade; and it is a very reasonable anticipation, consequent upon present developments, with steadily advancing yields, that, in a few years, it will present a city of metropolitan appearance.

The bridge which connects the two parts of the city is a wooden structure, and over it heavy teams are continually crossing, with a large number of vehicles and horsemen. It commands a view of the mouth of Oil Creek with its multitude of barges, while a view up the creek reveals a fleet of boats jammed to such a degree that a stranger may well wonder how even the most skillful boatman could navigate through them. From this bridge the great Pond Fresh, of which a very correct view is given, was witnessed by crowds, who looked upon the grand spectacle with awe and wonder.

The average shipment from Oil Creek is estimated at about eight thousand barrels, which, at the present time, may be considered worth about forty millions annually. This, to a great extent, finds its way down the Alleghany River, where ready transhipment is made to the eastern cities. Of this large exportation Oil City enjoys a fair proportion. Some large refineries are located in and about the city, the business of which forms a heavy part of the capital employed in this emporium of the oil trade.

Oil Creek.

Oil Creek is the outlet of a small lake in the northwest part of Crawford County, between the villages of Riceville and Union. Its course is south-east; its length about thirty miles; and empties into the Alleghany one hundred and fifty miles above Pittsburg. From its source to its mouth explorations for oil have been made, and immense quantities obtained. Along its borders a forest of derricks may be seen, some producing flowing wells, while others are pumping.

The most productive wells are located within twelve miles of its mouth. However, many very remunerative wells are in operation above this distance, and the territory indicates oil, although perhaps not equally developed. The Watson flats, below Titusville, corroborate this by late developments in many good wells.

The average depth of wells along the creek is about seven hundred feet—the greatest depth, one thousand feet. The plats have yielded the richer products in the aggregate, yet many wells on the hill-sides have been very compensatory, among which we may mention the Noble Well, which flowed, at its first opening, about twenty-five hundred barrels daily. This well, with others, with their derricks and immense tanks, furnish a very picturesque sight to a stranger, as he meanders his way amid the tanks, the derricks, and the industrial efforts along the banks of the creek.

The plats very frequently alternate—now upon one side of the creek, again upon the other. Their width is from one-eighth to half a mile, and present a business appearance never before witnessed among the hills of Pennsylvania. By reference to the chart, two tributaries of very great importance may be traced. The positions along the creek are designated, as will be observed, by farms—as the Clapp farm, the McClintock farm, the Buchanan farm, and others. Between the Buchanan farm and the J. McClintock farm is the terminus of Cherry Run, a small tributary running almost parallel with the creek itself. Many explorations have been made on this run with abundant yields, some wells producing from two hundred and fifty to three hundred barrels daily. Not far from its source the pleasant village of Plumerville is located, whose citizens enjoy a church, school, hotel, and the usual surroundings of a pleasant village, near which place may be seen the refineries of Messrs. Warren & Co., and one under the name of the Humboldt Refinery, under the highly successful management of Messrs. Bucz, Ludovick & Co. Both refineries have large capitals employed, and are considered equal, in wealth and immensity of business, to any other in the oil regions.

It will be observed that Cherry-tree Run, which rises between the Rynd and the Dempsey farms, is another tributary of Oil Creek. Along its banks most valuable territory may be found. As yet, it has scarcely received that attention which other departments of this region have.

The Oil Creek region possesses almost fabulous wealth, both in a state developed and undeveloped, and invites the attention of capitalists from Europe, as well as from all districts of the United States, and may now be considered, in point of present and prospective wealth, the India of America.

WARREN.

Warren, the county seat of Warren, is situated on the right bank of the Allegheny. It is elegibly situated, and commands picturesque views both up and down the river. It was incorporated in 1832.

Near the center of the town, which is about one-fourth of a mile from the river, is the public square; around its sides are situated the public buildings. It contains an academy, which, with the court-house, is built of brick. The jail, prothonotary's office, and bank building are of stone, and of respectable architectural appearance.

The place contains five churches: the Methodist Episcopal, German Methodist, Presbyterian, Lutheran, and Roman Catholic. The dwellings and business houses are generally wooden, firmly erected, and well painted, the neatness and cleanly appearance of which strike a visitor's eye upon entering the town. However, many costly and elegant brick buildings have been erected within a few years past. We may enumerate the Carver House, the Tanner, the Watson, the Johnson blocks, besides many private residences. The seat of justice was fixed at Warren at the organization of the county, March 16, 1819. It is two hundred and twenty-three miles from Pittsburg by land, and two hundred and three by river.

The business of Warren varies with the seasons of the year. During the rafting freshets, the whole country, as well as Warren, presents a business-like appearance. All is bustle with the preparations of lumbermen.

Warren may be considered at the head of steam navigation on the Allegheny River, many mill-dams having been constructed up the river, which obstruct the further passage of steamboats. Heavily freighted keel-boats from Pittsburg also arrive and depart at certain seasons of the year. During these rafting seasons Warren enjoys a harvest of trade, to which it is so justly entitled, from the character and liberality of its business men.

TIDIOUTE.

Tidioute is situated on the right side of the river. It comprises two considerable villages—Upper and Lower Tidioute—which are about a mile apart. It contains a number of stores, taverns, mechanic shops, etc. It likewise has Presbyterian and Methodist churches.

Here we enter upon the oil territory of the Allegheny River, undeveloped, developing, and developed. Evidences of oil have long been observable along the river margins, and the enterprise of individual capitalists and companies of large capitals have ventured on this region. Some have been successful to a satisfactory extent, while others are busily engaged in erecting derricks, importing engines, together with other accompaniments necessary for well-boring. The evidences are very favorable. It may safely be predicted that this region will, at no very distant day, be able to compete with territories of older development. If enterprise, activity, and capital are true levers of success, the prediction seems safe, and need not be feared. The town is enlivened by the presence of oil operators, speculators, agents for Eastern companies, and others who are on prospecting tours.

"Petroleum is one of the newest of our products. It was discovered just when it was wanted—when there was trouble in the manufacturing world about the production of oils, of which machinery alone required immense quantities for lubricating purposes. The problem of a cheap light was also becoming serious, in consequence of the exhaustion of the whale fisheries, and even by the increasing value of the fat of the hog, in consequence of its employment for illumination. In camphene and such burning fluids there were dangerous gases, which led to frequent explosion of lamps. But another use of petroleum presents itself. At the commencement of the Southern Rebellion, the North was completely cut off from the turpentine districts. The very small supply in Northern hands quickly advanced to exorbitant rates. In fact, it could not supply the demand for painting and other mechanical purposes. But, happily, in the refining of crude oil, an article designated benzole was discovered, the chemical purposes of which proved themselves the equal, if not the superior, of the products of the Southern pine, at once relieving the wants of builders and other mechanics. It now forms an important article of domestic as well as foreign commerce. Such are the results of the late oil discoveries, administering to the wants of mankind in so numerous effectual ways.

"Petroleum was introduced to the world just at the moment when it was sorely needed. To our own people it has been a blessing, but to foreign nations a boon, which it is shown they appreciated, by the heavy and increasing demand for the article, which has already become an important commercial material."

During the rafting freshets Tidioute becomes lively in business aspects. There are several large saw-mills in its vicinity, many of which have temporary railroads to carry their lumber to the river. The lumbering business furnishes employment for several hundred men, adding much to the trade of the place.

FRANKLIN.

Franklin, the county seat of Venango County, Pennsylvania, is situated on the Allegheny, at the mouth of French Creek. It was laid out in 1795, and contains the usual county buildings, together with six churches. There are several large mills and furnaces in the vicinity, from which much trade is derived. The dams on French Creek afford immense water-power for mills. Population, 1,500.

The locality of Franklin is now considered one of the great reservoirs or basins of the oil regions. Its immense productive character, which is now being developed, proves the certainty of obtaining an abundant and steady flow of oil wherever the territory is pierced. Numerous oil-wells have been struck, and the indications are that the supply seems to be inexhaustible. Its uses are multiplying and certain. The Allegheny and its tributaries produce the illuminating oils, benzine, lubricating and paint oils, etc. Many who have been carefully prospecting this stream have decided on purchasing and boring at or near Franklin, and every thing indicates certain success.

KITTANNING.

Kittanning, the seat of justice for Armstrong County, is a very beautiful town. It is handsomely situated on the left bank of the river. The buildings are good, and generally built of brick; many are handsome, and display the taste and refinement of the proprietors. Four streets run parallel with the river, and which are crossed at right angles by eight other spacious ways. It was laid out in 1804; incorporated into a borough in 1821. Population, about 3,500.

Visitors are often induced to remain some time, to enjoy the surrounding picturesque scenery and the pure, exhilarating atmosphere which this location affords.

The place contains a very fine hotel, which visitors delight to eulogize, on account of its really pleasant appointments. Large chambers, handsomely furnished, with high ceilings, excellent ventilation, a table with all that an epicure could desire; and, above all, a clever landlord, in the person of Mr. Reynolds, (after whom the house is named, "*The*

Reynolds House"), are sufficient to seduce a weary traveler to "put up" for healthful invigoration and the enjoyment of a temporary home.

FREEPORT.

Freeport, situated on the right bank of the Allegheny, is a flourishing village, and, for its handsome location, is second to no other town or village on the river. It possesses many natural advantages. Buffalo Creek empties into the river at the lower end of the town; and at its mouth is a flourishing village, which is a popular rendezvous for boatmen during the seasons of flood and ice.

Freeport Island, at the upper part of the village, forms a fine eddy in front of the town—an excellent landing-place for rafts and boats.

The canal from Pittsburg to Philadephia passes directly through here. Four daily trains of cars, on the Allegheny Valley Railroad, pass by on the opposite side of the river.

The place contains many large stores and hotels, besides many mechanic shops, foundries, steam-mills, woolen factories, etc. It was incorporated on April 8, 1833, and now contains nearly 2,500 inhabitants. It is blessed with seven churches: the Methodist Episcopal, Baptist, Presbyterian, Lutheran, Seceder, Episcopal, and Roman Catholic. With these advantages Freeport may be considered a progressive town, fully "up to the age."

TARENTUM.

Tarentum, situated on the right shore, above the mouth of Bull Creek, is an exceedingly well-built town. It is twenty-one miles above Pittsburg by the canal, which passes through it. The locks of the canal afford an excellent water privilege. Several mills are situated in the vicinity, propelled both by water and steam, besides large salt-works and coal-mines. The place contains the usual number of churches, stores, shops, etc.

SHARPSBURG.

Sharpsburg is situated on the right bank of the Alleghany. Its appearance, in a business view, indicates it a thriving town. It has had a very rapid growth, and its increase of population, buildings, etc., anticipate a large town at an early period. It contains several extensive manufacturing establishments, rolling-mills, sash factory, steam and keelboat building, etc. It is five miles from Pittsburg by the river.

THE ALLEGHENY RIVER.

DIRECTIONS FOR MAP NO. 1.

MORRISON'S BARS.

[Warren, 1 mile—Pittsburg, 202 miles.]

AFTER passing the pier of a bridge, (that once spanned the Allegheny, and which, after being unroofed in a gale of wind, was suffered to become so much injured and exposed that it gave way, and its wooden remains floated down the river), the pilot should keep near the right shore as far as the head of the riffle, one-half a mile above the bars. When half-way down the riffle, incline nearly to the middle, so as to pass between the two bars. In high water, little or no notice need be taken of the right-hand bar.

MEAD'S ISLAND.

[Warren, 3¼ miles—Pittsburg, 199¾ miles.]

This island is large, containing three hundred acres, and very valuable. It is owned by Mrs. Mead.

The channel here is to the right or left, but the right side is the better. By this, a pilot need not cross back to the right again above Grass Flat Islands.

Mead's Bar, from the island, stretches about one-fourth of a mile below the head. It reaches nearly half-way from the island to the right shore, throwing a strong current into the right bank. Turn short around the bar, as chart directs. When past the foot of the island, keep near shore, and look out for

JACKSON'S AND GRASS FLAT ISLANDS.
[Warren, 4¾ miles—Pittsburg, 198½ miles.]

These islands have frequently shown themselves to be troublesome customers to many lumbermen. Thousands of dollars' worth of lumber have been lost upon them. Pilots should follow the channel to the extreme right of these islands, which is always the deepest and safest, until having passed the lower Grass Flat until even with the end of the latter island; then hold to the left, and prepare for

SCOTT'S ISLAND.
[Warren, 5 miles—Pittsburg, 197 miles.]

Here the channel is to the left. The island is close to the right shore. There is little water to the right of it. Scott's Eddy is at the extremity of the island.

BIG BROKENSTRAW CREEK.
[Warren, 6 miles—Pittsburg, 196 miles.]

This creek rises in Erie County, Penn., and after receiving the waters of Coffee, Hair, Spring, Mullengar, and the Little Brokenstraw Creeks, empties into the Alleghany, seven miles below Warren.

The Little Brokenstraw takes its rise in Chautauque County, not far from the head-waters of French Creek, and empties into the Big Brokenstraw, seven miles above its mouth. The lumbering business is extensively pursued on all the tributaries.

Several millions of feet of lumber, together with an immense amount of shingles, yearly leave this point, and find a market throughout the country.

Brokenstraw Eddy is a few rods below the mouth of this creek, upon the right side, forming a very excellent place for preparation for lumbermen.

The village and land for a considerable distance around is owned by Dr. Wm. A. Irvine, grandson of General Wm. Irvine, of the Revolutionary army, and who was for several years Commissioner of the State, in superintending the survey of lands northwest of the Alleghany, under a law of 1792.

BROKENSTRAW ISLAND.

[Warren, 7¾ miles—Pittsburg, 195¼ miles.]

Here we have the channel to the right. The island is about three-fourths of a mile long, and contains sixty-three acres of cultivated land.

From the head of the island, in low water, keep near the middle. When approaching Dr. Irvine's house, incline to the right, to avoid the bar on the left, about opposite the house. After passing the bar, keep a little to the right of the middle, as far as the extremity of the island. The water is always deeper along the right shore.

DIRECTIONS FOR MAP NO. 2.

J. THOMPSON'S ISLAND.

[Warren, 9¾ miles—Pittsburg, 193¼ miles.]

THIS island is owned by Mr. James Thompson. It is over a mile long, and contains fifty-seven acres of tillable land.

Here the channel is to the right. In low water, keep very near the right shore around the right point below Dunn's Eddy, until around to the mountain. You are then clear of the large bars which project from the head of the island. Keep near the right shore until Deerfield Bar has been passed, which is below the middle of the island, and reaches about half-way from the island to the right shore.

Look well to Deerfield Bar in all stages of water. After passing the foot of the island, keep the right shore down as far as

CLARKE'S ISLAND.
[Warren, 11½ miles—Pittsburg, 191¼ miles.]

Channel again to the right. Enter the chute about midway between the island and right shore. Incline slightly to the right to avoid a bar which projects out a short distance from the head of the island. Keep along the right shore until at the foot of the island. There is no channel to the left.

ROBERT THOMPSON'S ISLAND.
[Warren, 12¾ miles—Pittsburg, 190¼ miles.]

Here we have channels to the right and left. The main travel is to the right in all stages of water. From the foot of Clarke's Island, keep near the right shore, so as to be within a few feet of it while passing the head of the island. You are then clear of the large, flat bar which projects out from the head of the island, and reaches half-way to the right shore. After having passed the bar, run over to the left about half-way to the island.

Be careful when passing the foot of Clarke's Island; take care of the strong current that makes the left. Observe a large rock three rods from left shore.

STEWARD'S ISLANDS.

[Warren, 14½ miles—Pittsburg, 188½ miles.]

The first of these islands contains fifteen acres, and is owned by Mrs. Magee. The second contains twenty acres, and belongs to Peter Smith. Channel to the right. This is the first general crossing-place below Warren.

CHARLEY SMITH'S BARS.

[Warren, 16 miles—Pittsburg, 186 miles.]

The ice-gorge in 1853 nearly ruined this portion of the river for navigation in low water. There are four different channels, and not any of them deep, wide, and straight enough to run barges or rafts through with safety. The best of these, in a fair running stage of water, is the left-shore channel over Bar No. 2. This bar, having been formed by the ice, extends across the channel in a form known to rivermen as a "pocket," or "fish-basket." The left center channel, made by Bar No. 2. This channel is the deepest of the four by nearly a foot, but no large boat can turn in it.

The right center channel is between the tow-head and Bar No. 1. It is two inches deeper than the left channel on Bar No. 2. The fourth is to the right of the tow-heads. While passing the bar that extends from the tow-heads diagonally, keep very near the right shore. This channel is narrow and crooked, although some think it the best in low water.

DIRECTIONS FOR MAP NO. 3.

MILL-STONE ISLAND.

[Warren, 17 miles—Pittsburg, 185 miles.]

HERE the channel is to the left, about the center of the left chute.

GOOSE FLAT ISLAND.

[Warren, 18 miles—Pittsburg, 184 miles.]

Channel to the left. There is also a channel to the right; but it is not traveled much.

MAGEE'S BAR.

[Warren, 19 miles—Pittsburg, 183 miles.]

Channel to the right. After leaving the lower Goose Flat Island, keep near the middle. When passing the bar, be a little nearer to the right than left shore; after which incline to the left.

COURSON ISLANDS.

[Warren, 19¾ miles—Pittsburg, 183¼ miles.]

Channel to the left. No navigable channel to the right.

MAGUIRE'S BAR.

[Warren, 21 miles—Pittsburg, 181 miles.]

Here the channel is to the right. This bar is situated on the left, a few rods above the foot of the third Courson Island. It reaches about half-way from the left shore to the island. When passing it in low water, leave two-thirds of the river to the left.

Tidioute Island.

[Warren, 22 miles—Pittsburg, 181 miles.]

Pilots should, in order to have a correct knowledge of this place, thoroughly examine the chart. In high water, rafts that come from above the Courson Islands always go to the left; but the center, or channel which runs between the head of the island and the upper middle bar, crosses the boat channel, and runs to the right of the lower middle bar, is preferable in low water. The right or low-water boat channel turns to the right at the upper end of the eddy, and above the upper middle bar passes through the eddy and down the riffle about three rods from the right shore, crosses the center channel, and runs down close to the island, and to the left of the lower middle bar. The latter is deeper than either of the others, but too crooked at the heads of the upper and lower middle bars for rafts to get into. The bar on the right, below Gordon's Run, reaches over half-way to the foot of the island. After passing the island, cross to the right shore.

White Oak Island.

[Warren, 26 miles—Pittsburg, 177 miles.]

From the foot of Tidioute Island to White Oak there are no less than ten islands and bars, which give the river a rugged appearance. From the center to the left shore, islands, bars, and tow-heads spot the river; the channel is, however, always to the right. They are well defined on the chart.

The famed White Oak Chute, which is the main traveled path in high water, is to the right of the island. The channel to the left of White Oak is considered the best in low water, and many like it the better in high water.

To run through this channel, hold over to the left, close to Isaac Jones' Island, after passing the little side-bar near the head.

DIRECTIONS FOR MAP NO. 4.
VENANGO COUNTY, PENN.
HEMLOCK ISLANDS.
[Warren, 27¾ miles—Pittsburg, 175¼ miles.]

CHANNEL to the left. In low water keep near the shore while passing the head-bar of the island, then incline toward the middle, and, when near the foot of the island, keep slightly toward the right of the middle, as far as the foot of the third, or lower island, then steer a straight course down about the center of the river, between Prather's Bars, passing to the right of Prather's Island, thence to the left, when you approach the bend below. By this you avoid Siggin's Bar, which projects from the right shore, and reaches half-way across to Dorkaway Island. When the river is very low, run quite near to this island.

In rafting stages, Fishing Bar is under water. Be not deceived as to its precise location.

Flat-boats, in very low water, should keep near the left shore when passing Green's Landing. When about the middle of the last Hemlock Islands, turn short to the right.

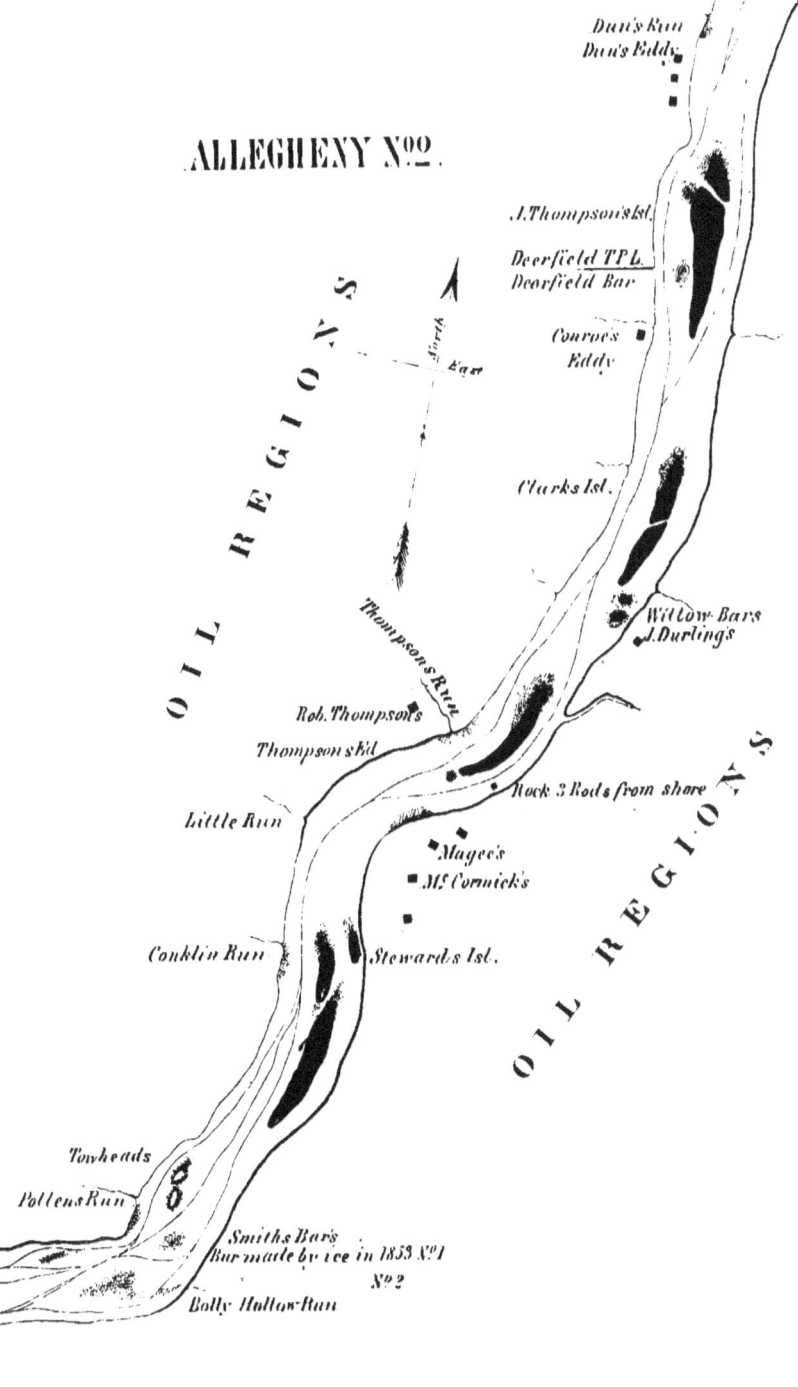

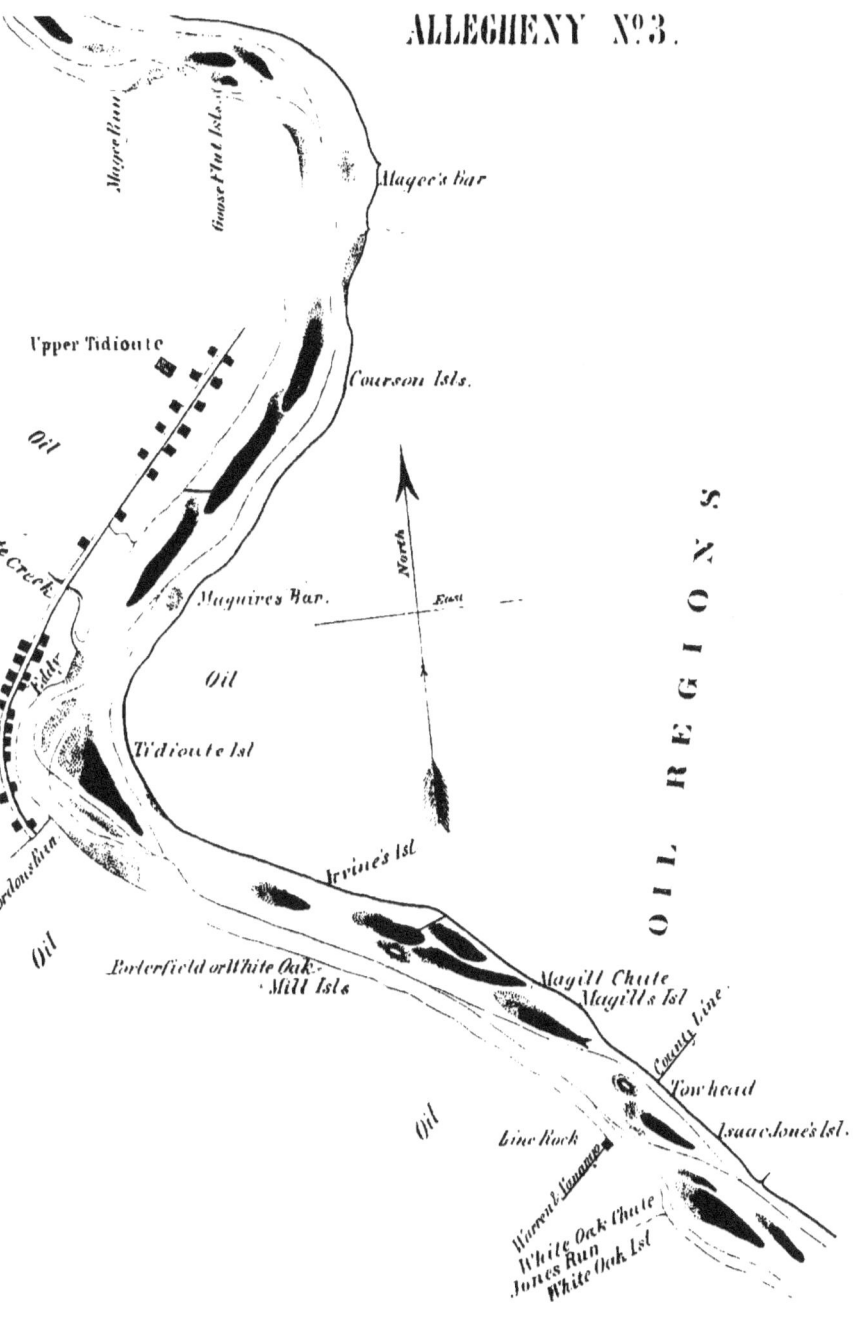

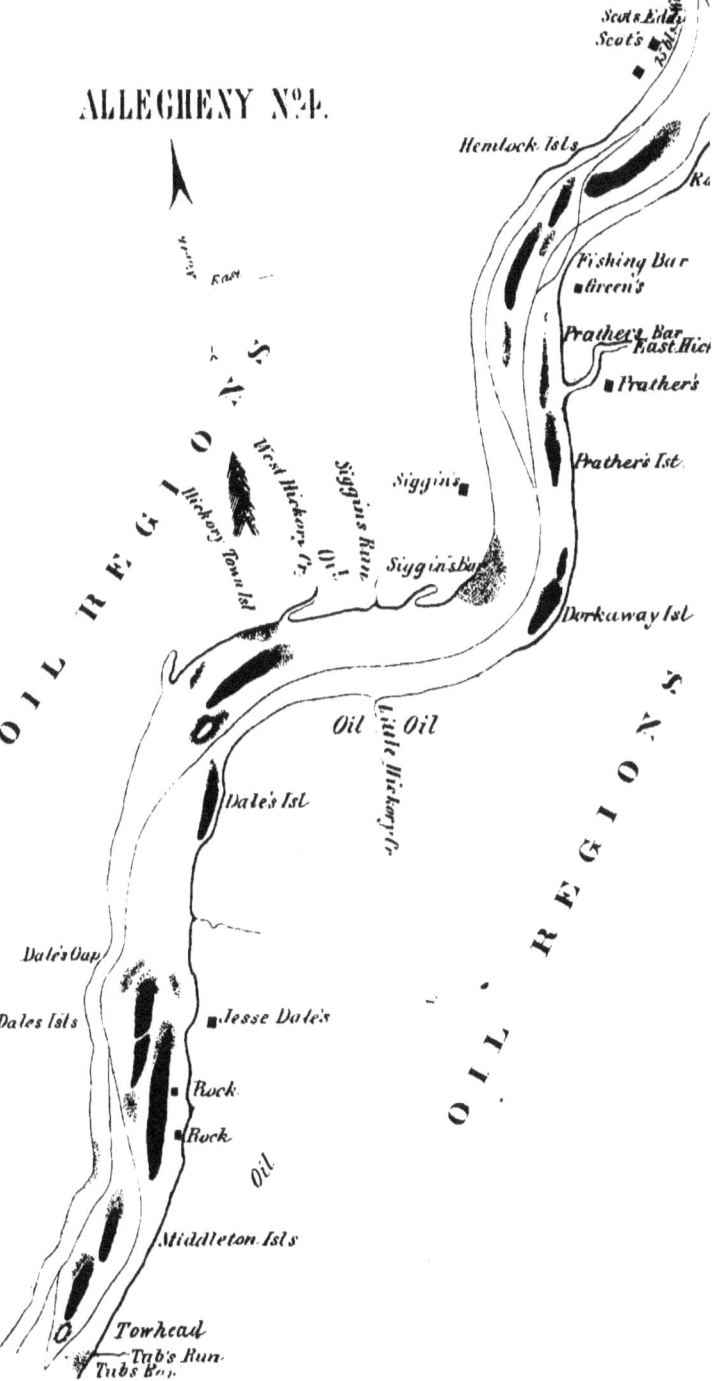

THE ALLEGHENY RIVER. 33

HICKORY TOWN ISLAND.
[Warren, 31½ miles—Pittsburg, 171¼ miles.]

Channel to the left. Keep slightly toward the left of the middle until the large tow-head at the foot of the island has been passed. Then make a long crossing to the right shore, to prepare for Dale's Gap.

The channel to the right of the island is very shallow, and frequently dry.

DALE'S ISLAND AND GAP.
[Warren, 33¼ miles—Pittsburg, 169¾ miles.]

Channel to the right. Keep quite near the right shore, around the right point, when entering the gap. After passing the large bar near the head of the island, keep slightly to the right of the middle until you arrive at the extremity of the island; then turn short around the right-shore point, in order to pass the bar of Middleton's First Island, close by. Keep slightly toward the right of the middle until about half-way down the second island; you are then clear of the head bar. Now turn to the left, so as to be close to the tow-head when passing it, at the foot of the island. After passing it, work over to the left, so as to pass about midway between Hunter's and May's Islands. Now turn slightly to the left, when descending the riffle, to avoid a large bar on the right, below the foot of Hunter's Island. Then go over to the right shore, to prepare for the Thomas Islands.

The channel to the left of Middleton Islands is much deeper to the right, and better in low water. To run through this channel, you should, when approaching the foot of Dale's Island, keep gradually to the left, so as to be quite near the gravel bar at the foot of the island. After passing the bar keep to the left, and pass about midway between the foot of Dale's left-hand Island and the head of Middleton's Upper Island.

After passing the head of second Middleton Island, incline to the right, and run quite near the tow-head at the foot. You then avoid Tub's Bar, which makes out from the left at the mouth of Tub's Run.

DIRECTIONS FOR MAP NO. 5.
TIONESTA ISLANDS.
[Warren, 37½ miles—Pittsburg, 165½ miles.]

These islands are thirteen in number, and extend along the river for about two miles. Some of them are under a high state of cultivation.

Channel to the right. After passing the head of the second island, keep near the middle until about half-way down, then gradually work over to the left. Having passed the head of the lower island, keep to the left, so as to run close to its foot. This will carry you clear of Cushon's Bars on the right, which extend nearly to the foot of the island. Now cross directly to the left shore; and, when crossing, take care lest you be driven upon Cushon's Bars by the current that comes down from the left of the island.

Holman's Eddy, a short distance below these islands, is a good landing.

HOLMAN'S ISLAND.
[Warren, 41 miles—Pittsburg, 162 miles.]

Here the channel is to the left. About half-way down the island are Holman's Bars.

In high water, keep close to the left shore, and to the left of the bars; but in low water the better channel is to the right of the bars. When even with the head of the island, keep slightly to the right of the middle. After passing the bars, turn back to the left, so as to be above the head bar of Holman's Second Island. The water is not as deep, by six inches, to the left of Holman's Bars as at Maple Island.

Maple Islands.

[Warren, 42 miles—Pittsburg, 161 miles.]

Channel about midway between the two islands that are on the right, and one close in to the left shore. When passing the lowermost island, keep about the middle of the river, and also while approaching the bars in the bend immediately below. When around the bend, cross to the right shore, to prepare for Hemlock Creek Islands.

DIRECTIONS FOR MAP NO. 6.

Hemlock Creek Islands.

[Warren, 44½ miles—Pittsburg, 158½ miles.]

Channel to the right. Keep close to the right shore until you have passed all of the islands; after which cross to the left. This channel is a little deeper than either of the others. Flat-boats can go to the left of the first two islands, and to the right of the last two. There is also a center channel, which is frequently navigated with boats. This is between the first two, and to the right of the last two.

Muskrat Eddy, a short distance below, is a good landing.

McCray's Island.

[Warren, 46¼ miles—Pittsburg, 150¾ miles.]

Main channel is to the left. From the foot of the Hemlock Creek Islands, keep near the left shore until the tow-head has been passed. Henry's Bar is situated in the bend below, a little to the right of the middle, and opposite Henry's residence. Channel to the left.

Pithole Island and Bars.

[Warren, 48¾ miles—Pittsburg, 154¼ miles.]

After passing Henry's Bar, cross to the right. When going down Pithole Riffle, keep quite near the right shore, to avoid a rough, rocky bottom on the left.

Walnut Islands.

[Warren, 50¼ miles—Pittsburg, 154⅜ miles.]

The low-water channel is to the right; but, in fair running stages, the left is preferable. It is not so crooked; besides, the distance is less. When approaching the island on the left above, observe the strong current which runs toward the right of the island.

Downing's Bar and Horse Creek Eddy.

[Warren, 53 miles—Pittsburg, 150 miles.]

This bar is so situated as to turn the main current of the river directly to the left shore. It, therefore, aids in forming Horse Creek Eddy, which lies immediately below the bar on the right, and is the usual landing-place on the eve of the first day's run from Warren.

In order to pass to the left of the bar and land in the eddy, keep down about the middle of the river until you have passed the head of the bar, then turn short to the right. This is the deepest and safest channel. In very high water, rafts can go to the right of the bar, by keeping close to the right shore when running around the right point above, and land in the eddy with comparatively little labor. Never attempt this in low running stages. Sometimes boats and rafts land on the left, near the Fuman.

DIRECTIONS FOR MAP NO. 7.

Horse Creek Island.

[Warren, 54¼ miles—Pittsburg, 148¾ miles.]

CHANNEL to the right. After passing the foot of the island, keep nearer to the right shore until you have passed Alcorn's Bars, which are to the left of the middle, the lowest of which is situated in the bend below.

Oil Creek Island, No. 1.

[Warren, 57½ miles—Pittsburg, 145½ miles.]

Channel to the right or left. In going to the right, after passing the head of the island, run closely to it, to avoid the large bar on the right below the mouth of Oil Creek.

Oil Creek Eddy commences at the foot of the bar. This is the place for the embarkation of the immense yield of petroleum from Oil Creek. Oil City is situated at the mouth of Oil Creek, and extends down the river as far as Thomas Morran's. [*See Articles on " Oil City, Oil Creek, and Oil Regions," pages* **14–24**.]

Oil Creek Island, No. 2.

[Warren, 58¼ miles—Pittsburg, 144½ miles.]

Channel to the right. From the foot of the island keep about three-fourths of the river to the left until you have passed nearly around the bend below, which is about one mile, after which incline over to the left shore. You are then clear of Holiday's Bars, at the head of the bend, on the left.

Shafer's Island.

[Warren, 61¼ miles—Pittsburg, 141¾ miles.]

Channel to the left. Keep slightly to the left of the middle of the chute when passing the head bar of the island.

DIRECTIONS FOR MAP NO. 8.

Two-Mile Run Island.

[Warren, 62½ miles—Pittsburg, 140½ miles.]

Channel to the left—slightly to the left of the middle of the chute.

McDowell's Island.

Channel to the left. Keep about the middle until at the foot of the island. You are now about the right distance from the left shore to run around the bend, and pass through the third space from the right shore to the remains of Franklin Bridge.

In order to land in Franklin for the purpose of doing business, it will be necessary to take this channel, and cross the mouth of French Creek.

In high water, boats can be landed any-where between the bridge and lock. There is always plenty of water on Old Garrison Bar; but in low water it will be necessary to land either below the bar, and a little above the bridge, or at the upper steamboat landing, a few rods below the lock. The upper landing is performed by towing up, either before or after crossing French Creek. [*See article on "Franklin," page* 21.]

HANGING ROCK BAR AND RIFFLE.
[Warren, 76¼ miles—Pittsburg, 126¼ miles.]

Channel to the right or left. The easier and better channel is on the left. When you are nearly at the mouth of Big Sandy Creek, which comes in on the right above incline toward the left shore. After passing the bar cross to the right. This channel should always be run, except when leaving Big Sandy Eddy; but with a large boat or raft it should not be attempted. After passing the bar, keep near the right shore down to

APPLEGATE'S RIFFLE AND WITHERUP'S BARS.
[Warren, 77¼ miles—Pittsburg, 125¾ miles.]

Channel to the right. Keep near the right shore while going down the riffle; and, when down to the foot of the bars, prepare to go either to the right or left of Steen's Island.

STEEN'S ISLAND.
[Warren, 78¾ miles—Pittsburg, 124¼ miles.]

Channel to the right or left. The deepest water is on the right; but the main track is to the left in almost every stage of water. The water is about as deep on the left as it is on Applegate's or Charley's Riffles. After passing the island, keep near the right shore, to prepare for

CHARLEY'S RIFFLE AND BARS.
[Warren, 80 miles—Pittsburg, 123 miles.]

Channel to the right. Keep near the right shore while going down the riffle, and when around the bend incline over to the left, to prepare for Williams' Bars.

DIRECTIONS FOR MAP NO. 9.

WILLIAMS' BARS.
[Warren, 81¾ miles—Pittsburg, 121¼ miles.]

CHANNEL to the left. These bars are situated about three-fourths of a mile above the mouth of Denison's Run, which approaches from the right. When passing them, keep about three-fourths of the river to the right. After passing them, incline to the right, to prepare for

BIG SCRUBGRASS ISLAND AND BARS.
[Warren, 83¼ miles—Pittsburg, 119¼ miles.]

Here we have three channels, all of which are to the right of the island. The one generally used, and probably, under all circumstances, the safest, is the right-shore channel, and to the right of Pilot Rock. This channel is rather difficult to navigate, in consequence of a heavy press of water toward the right shore below Pilot Rock.

The second or middle channel is between the Pilot Rock and the tow-head. It is the deepest, and requires skill to navigate it. To go through it with

ALLEGHENY No 5

Map features labeled:
- Tubbs Run
- Tionesta
- Big Tionesta Cr
- Oil
- Lit Tionesta Cr
- Cushon Bars
- Hollman's Eddy
- Hunters Run
- Pierces Run
- Panther Run
- Hollmans Ist
- Hollman's
- Hollmans Bars
- Hollmans lower t
- Maple Is.
- OIL REGIONS

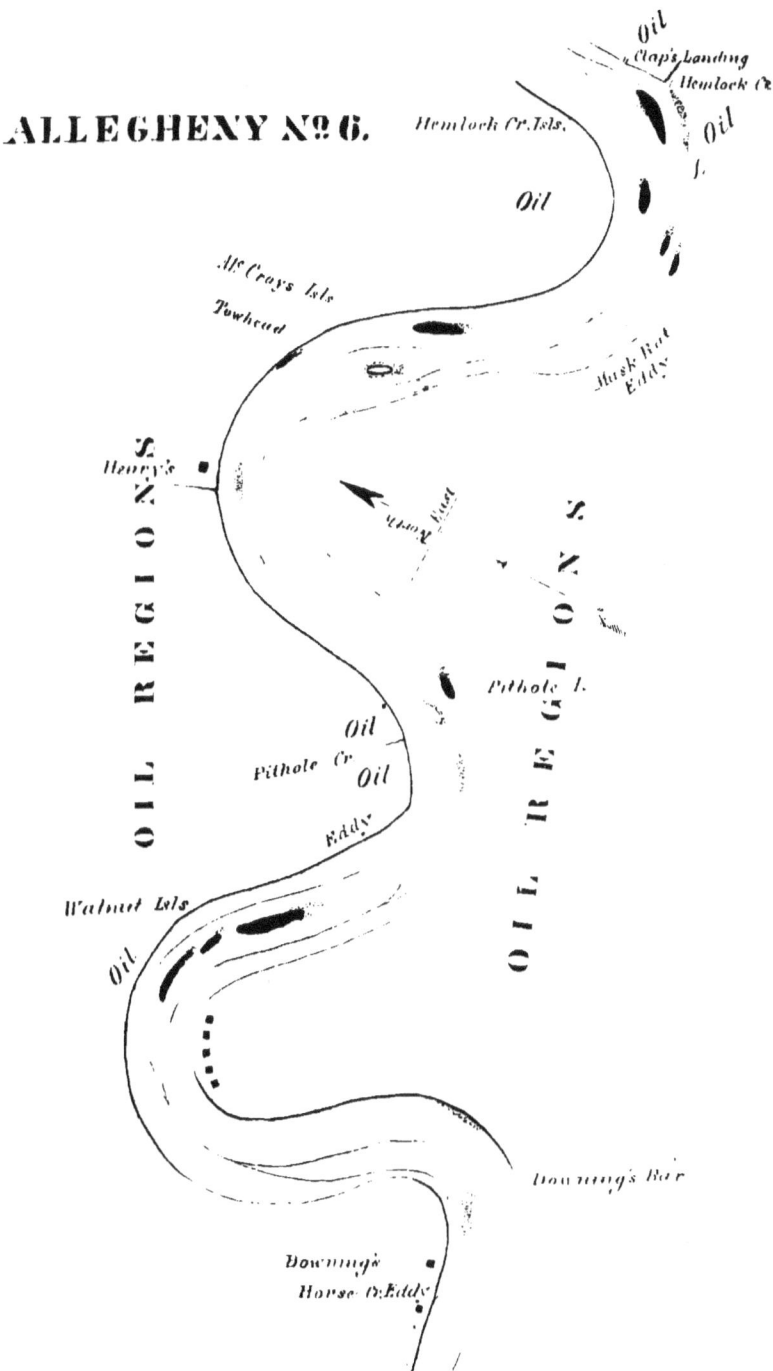

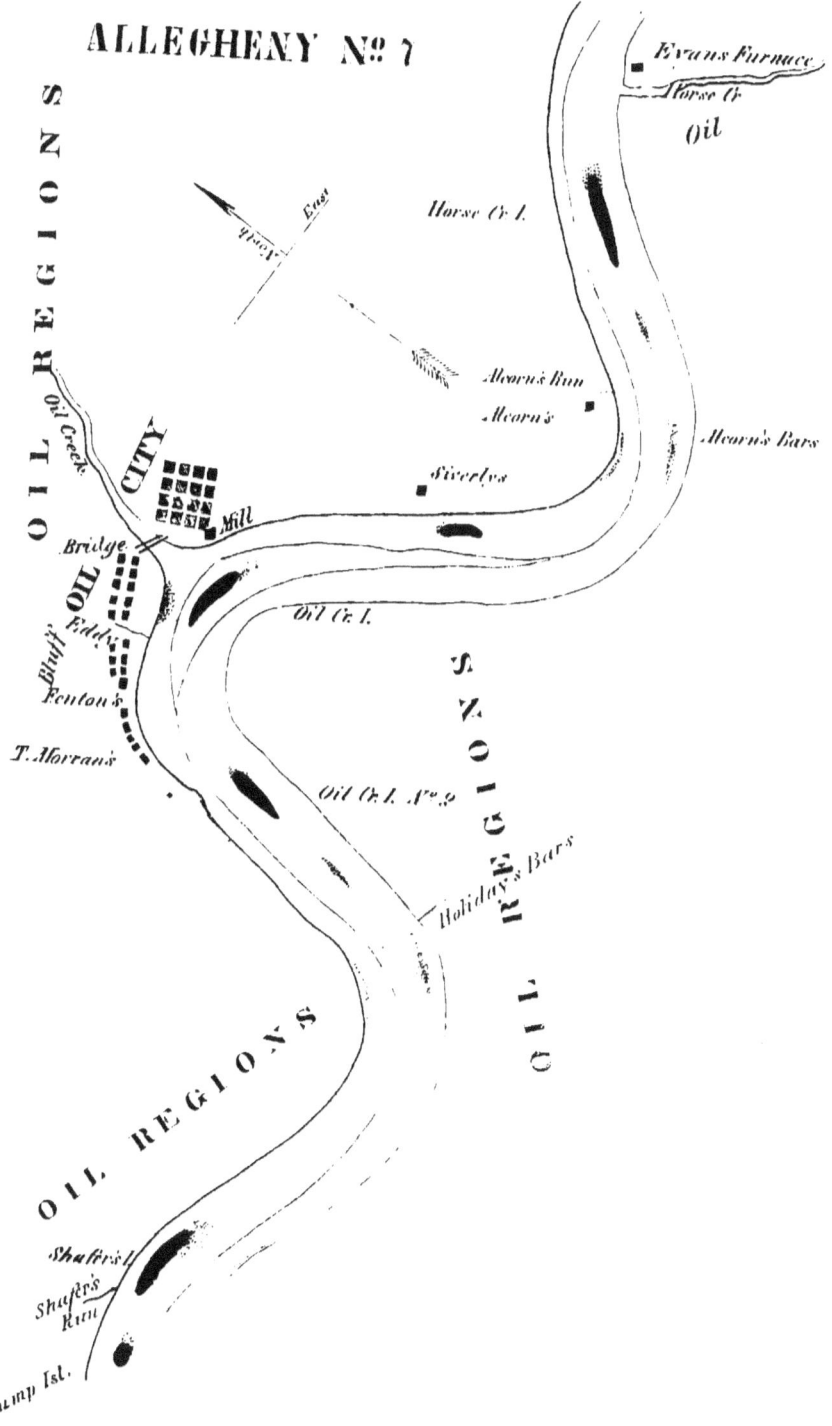

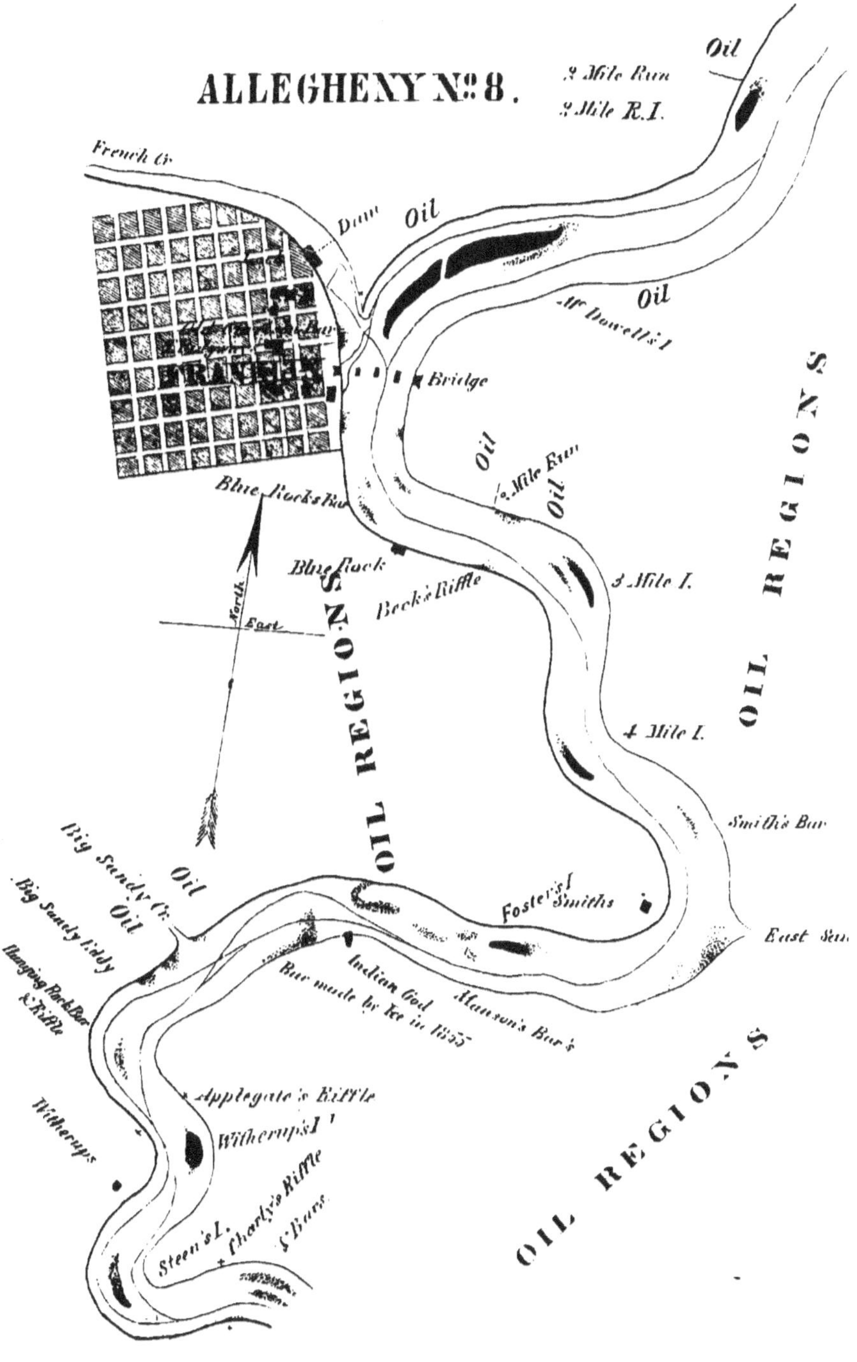
ALLEGHENY No. 8.
Oil
3 Mile Run
3 Mile R.I.
French Cr.
Dam
Oil
Oil
Mc Dowell's I.
Bridge
Oil
Oil
Blue Rocks Bar
Blue Rock
Beck's Riffle
3 Mile I.
North
East
OIL REGIONS
OIL REGIONS
4 Mile I.
Smith's Bar
Big Sandy Cr.
Oil
Oil
Big Sandy Eddy
Foster's I.
Smiths
East San
Hanging Rock Bar & Riffle
Indian God
Bar made by Ice in 1855
Hanson's Bar's
Applegate's Riffle
Witherups I.
Witherups
Charly's Riffle
Steen's I.
S. Bars
OIL REGIONS

a large boat or raft, keep slightly toward the right of the middle of the river. Be careful not to run upon the Pilot Rock or the tow-head on the left.

The third channel between the tow-head and the foot of the island is not safe for large rafts, but good for flat-boats.

After passing these bars, keep near the right shore, to prepare for

JACOB'S BARS.

[Warren, 85½ miles—Pittsburg, 117½ miles.]

The deepest and best channel is to the right of Bars No. 1 and No. 2. Cross to the left to avoid Bar No. 3, which juts out from the right shore nearly around the bend. After this, cross back to the right, to avoid Bar No. 4.

The extreme left channel is a narrow one, excavated by keel-boatmen.

JACOB'S, OR FALLING SPRING RIFFLE.

[Warren, 86½ miles—Pittsburg, 116½ miles.]

Eddy on the left, below the riffle. Falling Spring on the right.

JACOB'S EDDY, ON THE RIGHT.

[Warren, 88¾ miles—Pittsburg, 114¼ miles.]

Here we have a large bar on the eddy. Rafts and boats frequently tie up for the night directly over it, although not very safely, especially when the water is falling fast.

Montgomery's Falls and Elephant Bars.

[Warren, 90½ miles—Pittsburg, 112¾ miles.]

Here we have again three channels. In low water, while passing the Elephant Bars at the upper falls, keep by the right-shore channel. When you are at the foot of the riffle, incline to the left and run down about the middle of the lower falls.

In high water, keep close to the left shore while passing the bars; afterward incline toward the middle of the river.

The center channel is wider and a little deeper than the left channel.

Davis' Bar.

[Warren, 91¾ miles—Pittsburg, 111¼ miles.]

This bar makes out from the right-shore point in the bend, and reaches across the river more than half-way. In good running stages it is not noticed. Steer slightly toward the left of the middle when passing it; but in low water run around near the left shore.

Craig's Eddy is just below, on the left.

Evault's Defeat Island.

[Warren, 93½ miles—Pittsburg, 109¾ miles.]

Channel to the left. In low water, keep near the head of the island when about to pass it. By this you avoid a rough, rocky bottom along the left of the riffle. When you are two-thirds of the way down the riffle, gradually incline to the left. When the river is in good running stage, go down about the middle of the river. The right of the island is very frequently dry.

Stover's Island.

[Warren, 95¾ miles—Pittsburg, 107¼ miles.]

Channel to the right. This island is situated quite near the left shore. When at the foot of the island, keep about two-thirds of the river to the left until you pass Stover's Bar, immediately below the island.

Patterson's Falls.

[Warren, 97¾ miles—Pittsburg, 105¼ miles.]

Steer down the middle of the river. When you arrive at the foot of the falls, keep near the right shore, in order to pass the right of Goff's Bar, which is only a short distance below. This bar is situated in the middle of the river, and a short distance above a small stony point on the right shore. At low stages of water it is slightly covered with water. Look out for it.

When going around the bend below, incline to the right shore, and prepare for

Nicholson's Eddy and Hackney's Bar.

[Warren, 99¾ miles—Pittsburg, 103¼ miles.]

Channel to the right. When opposite the little right-shore bar, a few rods below the boat-scaffold, turn slightly to the right, in order to pass the head of Hackney's Bar; then incline more to the left. The head of the bar approaches so near the smaller one on the right, below the boat-scaffold, as only to leave a very narrow passage, barely sufficient to navigate through. This must be done in season, for the main current of the river at this point crosses directly to left of the bar; it then spreads over the bar, for the space of half a mile, back into the right channel. The bar is about a mile long. When you approach the foot, cross directly to the left shore, to avoid a rocky bar on the right point.

DIRECTIONS FOR MAP NO. 10.

Emlenton and Bridge.

[Warren, 102¾ miles—Pittsburg, 100¼ miles.]

The bridge has only one pier, which stands in the middle of the river. Channel to the right. In going past Emlenton, it will require more or less care to avoid rubbing the rocks along the right shore, until you approach Rocky Point, which is below the town, at the head of Ritchie's Riffle.

Emlenton, situated on the left bank of the river, presents a business-like appearance. It contains the usual supply of mechanic-shops, stores, etc.

Large quantities of grain are annually shipped from this place. Upon the whole, this is quite a point for trade.

Cumming's Trunk Riffle and Crawford's Bar.

[Warren, 104¾ miles—Pittsburg, 98¼ miles.]

Channel to the right. Keep slightly to the right of the middle while passing the bar, which lies quite near the left shore.

Stump Creek Eddy, on the Left.

[Warren, 106¾ miles—Pittsburg, 96¼ miles.]

This, or Miller's Eddy, is the usual place of landing, on the second night after leaving Warren.

Stump Creek Islands.

[Warren, 108¾ miles—Pittsburg, 94⅛ miles.]

Channel to the left. In very low water, after passing the head of the bar to the second islands, keep slightly out from the left shore, to avoid the rocky bottom on the left above the mouth of Clarion River. Now, turn short into the mouth of the creek; then out again; afterward hug the left shore until the middle of the third island has been passed, from which a bar juts out half-way to the left shore.

The low-water boat channel is to the right of the first two islands. Cross back again into the left channel, near to the foot of the second island. This channel is dangerous for rafts.

In high water, flat-boats can go to the right of the third island.

Parker's Landing is on the right; Graham's on the left, immediately below this island.

Parker's Bars.

[Warren, 110¼ miles—Pittsburg, 92⅜ miles.]

The main rafting track, in a good stage of water, is upon the left, about midway between the bar and the left shore.

The center channel is the deeper and better in low water.

Flat-boats can go to the right of the little bar opposite the mouth of Parker's Run, a short distance below Parker's Landing.

Parker's Falls.

[Warren, 111⅛ miles—Pittsburg, 91¾ miles.]

Channel about the middle.

Rattlesnake Falls.

[Warren, 112¼ miles—Pittsburg, 90¼ miles.]

Bar on the right and left. Channel about the middle.

DIRECTIONS FOR MAP NO. 11.

Miller's Eddy, on the Right.

[Warren, 115¼ miles—Pittsburg, 87¾ miles.]

This excellent eddy is formed by a large bar that juts out from the right shore, and turns the current toward the middle of the river. In high water it is two days' run from Warren. This is also a considerable place—business-like in its exterior. Mr. R. Criswell is heavily engaged in the manufacture of fire-brick, which forms an important item of export from this eddy.

Black Fox Island.

[Warren, 116 miles—Pittsburg, 87 miles.]

Channel to the right. In low water, keep toward the island, until the bar has been passed on the right, below the saw-mill. Now turn to the right, to avoid the large bar which projects from the head of the island.

Bald Eagle Island.

[Warren, 117¼ miles—Pittsburg, 85¾ miles.]

Channel to the right.

Armstrong's Rapids, or Truby's Riffle, and Truby's Bars.

In good running stages, after rounding Cinder Bank Bend, keep down the left shore. Some rocks in the water about three rods from the left shore, at the head of the riffle.

In very low water, steer down about the middle of the river, and pass between the middle bar and the long, flat shore-bar on the right. Now turn slightly to the right when going between the bars. Good Eddy on the right, in the bend below.

CATFISH FALLS.
[Warren, 122¼ miles—Pittsburg, 80¾ miles.]

Keep about two-thirds of the river to the left, at the commencement of the riffle, in order to pass between the middle bar and the mill-dam on the right.

The low-water channel for boats is down the left shore, and to the left of the middle bar, and very near the mouth of Catfish Run.

Catfish Eddy, on the left, below the large shore bar, and in front of the brick tavern, is a good landing-place for craft of all kinds.

At the foot of this eddy is Brady's Bend. Around the bend, on the left, a short distance below the coal-scaffold, under a point of rocks, is a good landing-place, and easy to land.

SUGAR CREEK RAPIDS AND BARS.
[Warren, 124½ miles—Pittsburg, 78½ miles.]

Keep down the left shore.

GOOSE BAR.
[Warren, 127¼ miles—Pittsburg 75¾ miles.]

Channel to the left. The main portion of this bar is situated in the short turn, opposite the point of the mountain where it comes down to the river on the left, below Snyder's coal-scaffold. The head of the bar extends diagonally toward, and nearly to, the coal-scaffold.

In order to pass this bar systematically, when around to the crossing-point rocks on the right about half a mile above the coal-scaffold, commence working gradually over to the left. The crossing is easy. The crossing-point rocks may be known by a thick grove of hemlocks back of them.

McClure's, or Magonigle's Bars.

[Warren, 128¾ miles—Pittsburg, 74¼ miles.]

Channel to the left. In low water, be close to the left shore, when opposite a house standing alone, about half a mile below the town of Phillipsburg. In good running stages of water, keep down the right shore, till the bars have been passed. Now, incline to the left, to avoid a rocky bar below the mouth of Frazier's Run. Below the rocky bar, on the right, is Gillespie's Eddy. Red Bank Eddy is on the left.

Red Bank Rapids and McClatchie's Bars.

[Warren, 131¼ miles—Pittsburg, 71¾ miles.]

Channel to the right or left. In low water, after leaving Red Bank Eddy, work gradually over to the right-shore point, opposite the mouth of Red Bank Creek. Keep near the right-shore point; then wear out toward the middle, to avoid a bar on the right, below the point.

In good running stages, keep down to the left of the long middle bar, about five rods from the left shore.

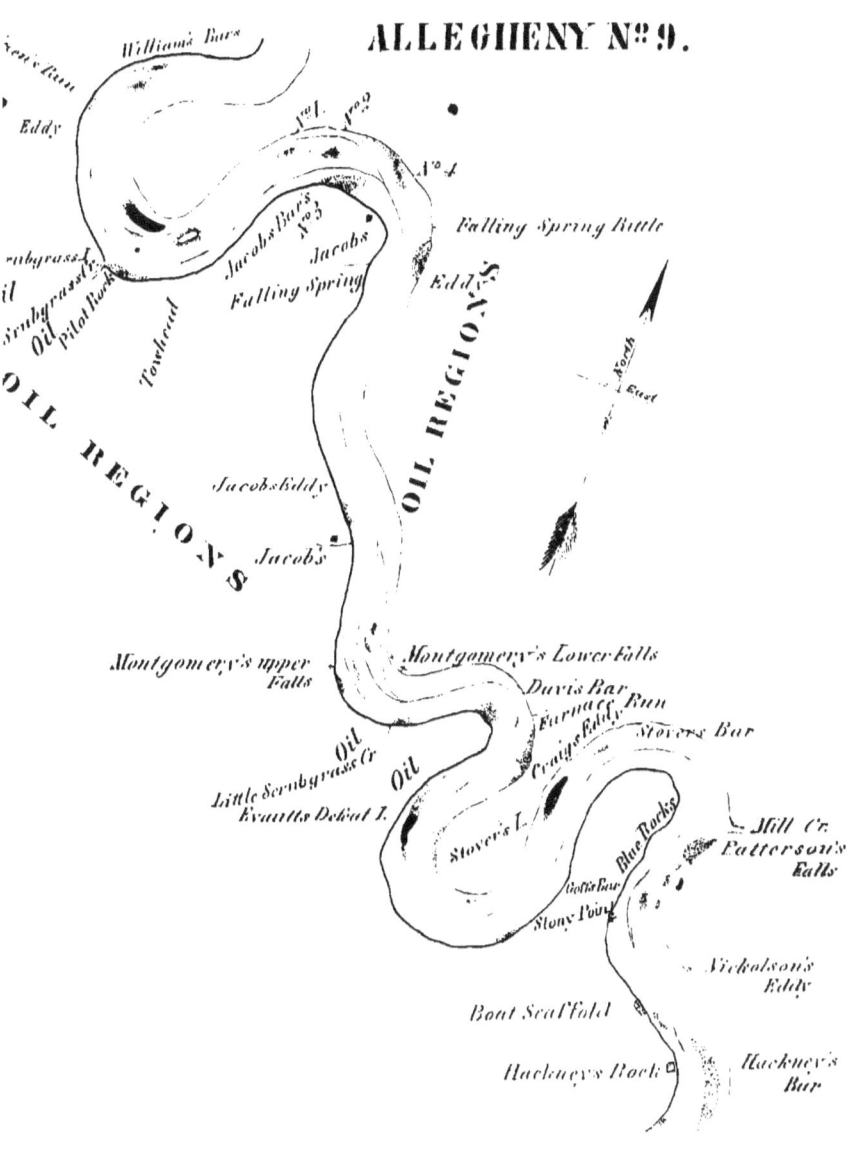

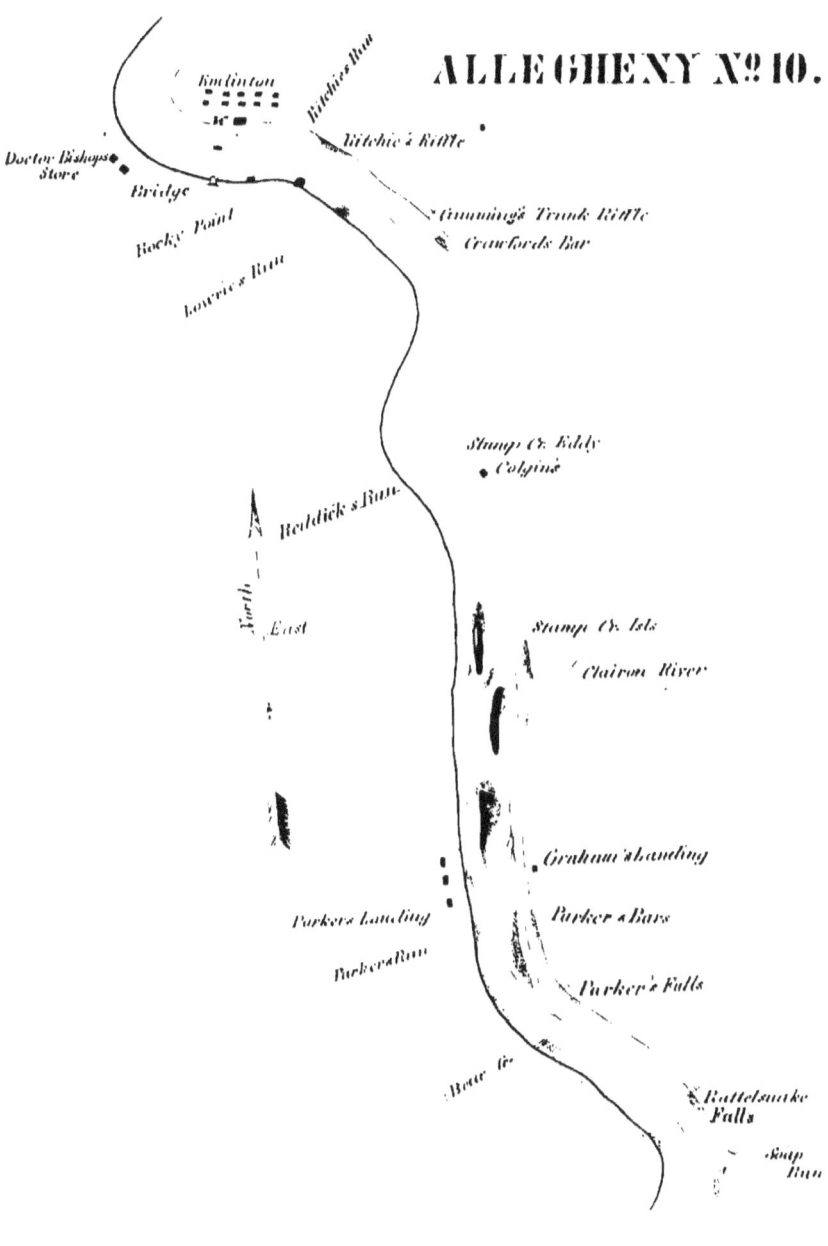

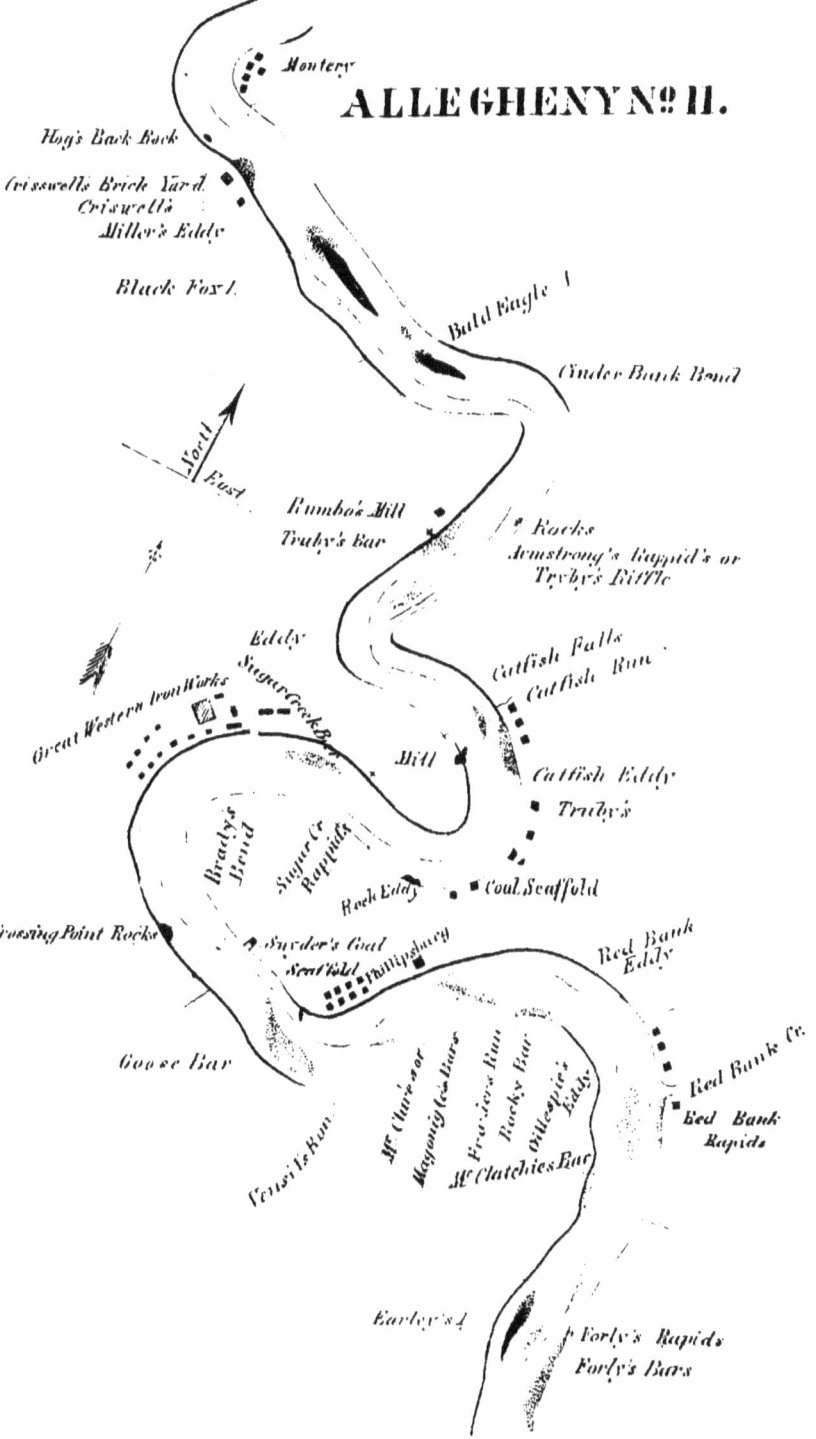

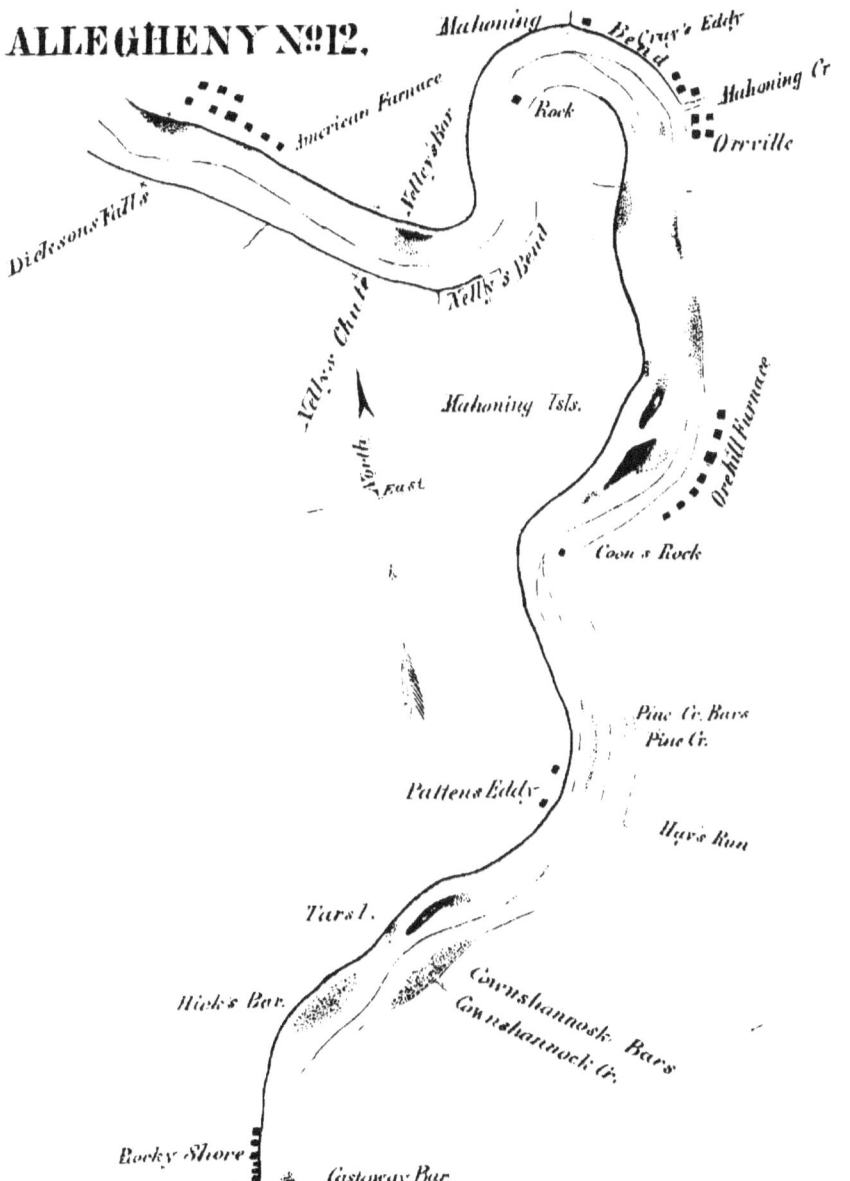

THE ALLEGHENY RIVER. 49

Early's Island and Farly's Rapids.
[Warren, 133½ miles—Pittsburg, 69½ miles.]

Channel to the left. In low water, be quite near the left shore, when passing the head of the island, in order to avoid the little middle bar. Then turn out to the right toward the foot of the island, to avoid the large bar that projects from the left shore.

DIRECTIONS FOR MAP NO. 12.

Dickson's Falls.
[Warren, 136 miles—Pittsburg, 67 miles.]

In very low water, the channel is near the right shore, at the head of the falls. The American Furnace is on the left.

Nelly's Chute and Bar.
[Warren, 138½ miles—Pittsburg, 64½ miles.]

Channel to the right. In low water, keep near the right shore, when passing the head of the bar.

The rock in the water near the right point, about a mile below, is of little account, except in very low stages of water.

Mahoning Rapids.
[Warren, 142¼ miles—Pittsburg, 60¾ miles.]

In low water, after leaving Gray's Eddy, work gradually across to the right, so as to be about two-thirds across the river. When opposite the mouth of Mahoning Creek, pass between two small stony bars, the left one being about the middle of the river. When the bars have been passed, incline to the left,

5

DIRECTIONS FOR MAP NO. 3.

MILL-STONE ISLAND.

[Warren, 17 miles—Pittsburg, 185 miles.]

HERE the channel is to the left, about the center of the left chute.

GOOSE FLAT ISLAND.

[Warren, 18 miles—Pittsburg, 184 miles.]

Channel to the left. There is also a channel to the right; but it is not traveled much.

MAGEE'S BAR.

[Warren, 19 miles—Pittsburg, 183 miles.]

Channel to the right. After leaving the lower Goose Flat Island, keep near the middle. When passing the bar, be a little nearer to the right than left shore; after which incline to the left.

COURSON ISLANDS.

[Warren, 19¾ miles—Pittsburg, 183¼ miles.]

Channel to the left. No navigable channel to the right.

MAGUIRE'S BAR.

[Warren, 21 miles—Pittsburg, 181 miles.]

Here the channel is to the right. This bar is situated on the left, a few rods above the foot of the third Courson Island. It reaches about half-way from the left shore to the island. When passing it in low water, leave two-thirds of the river to the left.

Tidioute Island.

[Warren, 22 miles—Pittsburg, 181 miles.]

Pilots should, in order to have a correct knowledge of this place, thoroughly examine the chart. In high water, rafts that come from above the Courson Islands always go to the left; but the center, or channel which runs between the head of the island and the upper middle bar, crosses the boat channel, and runs to the right of the lower middle bar, is preferable in low water. The right or low-water boat channel turns to the right at the upper end of the eddy, and above the upper middle bar passes through the eddy and down the riffle about three rods from the right shore, crosses the center channel, and runs down close to the island, and to the left of the lower middle bar. The latter is deeper than either of the others, but too crooked at the heads of the upper and lower middle bars for rafts to get into. The bar on the right, below Gordon's Run, reaches over half-way to the foot of the island. After passing the island, cross to the right shore.

White Oak Island.

[Warren, 26 miles—Pittsburg, 177 miles.]

From the foot of Tidioute Island to White Oak there are no less than ten islands and bars, which give the river a rugged appearance. From the center to the left shore, islands, bars, and tow-heads spot the river; the channel is, however, always to the right. They are well defined on the chart.

The famed White Oak Chute, which is the main traveled path in high water, is to the right of the island. The channel to the left of White Oak is con-

should not be done unless for the purpose of landing at Kittanning.

KITTANNING BRIDGE.

[Warren, 158¾ miles—Pittsburg, 49¼ miles.]

Channel first or second space from the right shore; second space is the deeper. After passing the bridge, keep near the right shore until you have passed the bar, situated about the middle of the river in front of the rolling-mill, at the lower end of the town.

In good running stages, flat-boats can go to the left of the bar.

Blue Rock Eddy is on the left, below the bar, and a short distance above the town of Manorville.

COGSLEY'S ISLAND.

[Warren, 155½ miles—Pittsburg, 47¾ miles.]

Channel to the left. After passing the bar at the head of the island, incline to the right, to avoid the bar on the left, opposite the foot of the island. When the island has been passed, cross over to the right, and keep near the right shore until Montgomery's Bars have been passed, and you approach

CROOKED CREEK ISLANDS.

[Warren, 158¾ miles—Pittsburg, 44¾ miles.]

Channel to the right. In low water, run around to the right of the middle bar, which lies even with the head of the second island.

At the foot of the bar, turn short to the left toward the island, to avoid being driven upon the bar that extends upward from the head of Sloan's Island, on the right. After this bar has been passed, incline to the middle of the river; make a long crossing toward the white rocks on the right shore, to prepare for Nicholson's Islands and Falls.

In good running stages, rafts run directly over the middle bar, to which we have already alluded, near the head of the second island.

Nicholson's Islands and Falls.

[Warren, 161¾ miles—Pittsburg, 41¼ miles.]

Channel to the right. In good running stages, steer slightly toward the right of the middle while going down the falls; but, in low water, after passing the head of the first island, incline to the left, and keep near the islands. After passing the bar, which is a short distance below the foot of the second island, keep near the middle of the river until you have passed the Cornfield Bars: the one on the left, which extends across the mouth of Taylor's Run; the other, on the opposite side, is near the town of Clinton.

Pickel's Eddy, on the Left.

[Warren, 164¾ miles—Pittsburg, 38¼ miles.]

This is the general landing-place on the eve of the third day's run from Warren. After leaving the eddy, steer so as to go to the right of Walker's Bar, a short distance below. The head of the bar extends up to the lower end of the eddy.

Murphy's Eddy is a mile below, on the right, and immediately below the cinder-bank to Hill's Saltworks, and opposite the foot of Walker's Bar. This eddy is convenient, but not so large as Pickel's.

DIRECTIONS FOR MAP NO. 14.

Murphy's Island.
[Warren, 166¾ miles—Pittsburg, 36¼ miles.]

CHANNEL to right or left. The rafting channel is to the right. About midway between the head bar of the island and right shore is a small lump, or bar; rafts can go either side of it.

In very low water, flat-boats should go to the left of the island, as it is the deepest water, and nothing in the way, except a small rock, which lies a little below the middle of and quite near the island.

Mad-Dog Island.
[Warren, 168¼ miles—Pittsburg, 34¾ miles.]

Channel to the right. After passing the head bar of the island, and when going down the riffle, care should be taken not to be driven against the right shore. The water presses hard to the right, and is very swift.

Aqueduct.
[Warren, 169¼ miles—Pittsburg, 33¼ miles.]

Channel first or second space from the right shore. Kiskiminitas Creek empties into the river on the left below the aqueduct.

In passing the railroad bridge below the aqueduct, always take the middle space. Look out for this place: it is well calculated to deceive. The bar on the right shore, a little below the aqueduct, turns the main current out toward the middle of the river. Many have attempted to descend through the next space to the right, but almost invariably shipwrecked on the second pier from the right shore.

KARN'S ISLAND.

[Warren, 173½ miles—Pittsburg, 29½ miles.]

Channel to the left. When going down the *reach*, work gradually over to the left; and, when passing the head bar of the island, be very near the left shore.

JACK'S ISLAND.

[Warren, 175 miles—Pittsburg, 28 miles.]

Channel to the right. While entering the chute, keep nearest the island, in consequence of a small bar near the right shore, and a small distance above the head of the island. When opposite the middle of the island, incline to the left, and pass near the foot, to avoid a large bar which juts out from the shore.

DIRECTIONS FOR MAP NO. 15.

BULL CREEK ISLAND.

[Warren, 178 miles—Pittsburg, 25 miles.]

CHANNEL to the left. About half a mile below the foot of the island, and near the mouth of Bull Creek, which enters upon the left, the Bull Creek bars commence. These bars extend diagonally, gradually nearing the left shore for about three-fourths of a mile. The end of the bars, and the narrowest place, is even with the lower salt-works. After passing the island, keep near the left shore, and when passing the lower salt-works, hug the left shore closely. In low water, be only a few feet from shore. After passing the bars, incline toward the middle.

PUCKERTY ISLAND.

[Warren, 183 miles—Pittsburg, 20 miles.]

Channel to the left. After passing the bar at the foot of the island, keep slightly toward the middle until a large bar on the left has been passed, which extends downward from the mouth of Poketus Creek. Logan's Eddy is on the left below the bar.

FOURTEEN-MILE ISLAND.

[Warren, 186¾ miles—Pittsburg, 16¼ miles.]

Channel to the left. In good running stages, keep the straight channel, and near the left shore, to avoid the large bars that make out from the island. In very low water, keep the left or crooked channel. [*See chart.*] The little bar near the white rock is of no consequence, except in very low running.

Huland's Eddy is on the right, below the mouth of Deer Creek.

NINE-MILE ISLAND.

[Warren, 191½ miles—Pittsburg, 11½ miles.]

Channel to the right or left. The best rafting channel is to the left, although it is not traveled, perhaps, as much as the right. The crossing to the left is made very easy, if commenced on the riffle, at Powers' Run, about a mile above. This channel is not crooked, and is much more easily navigated. [Study the chart.]

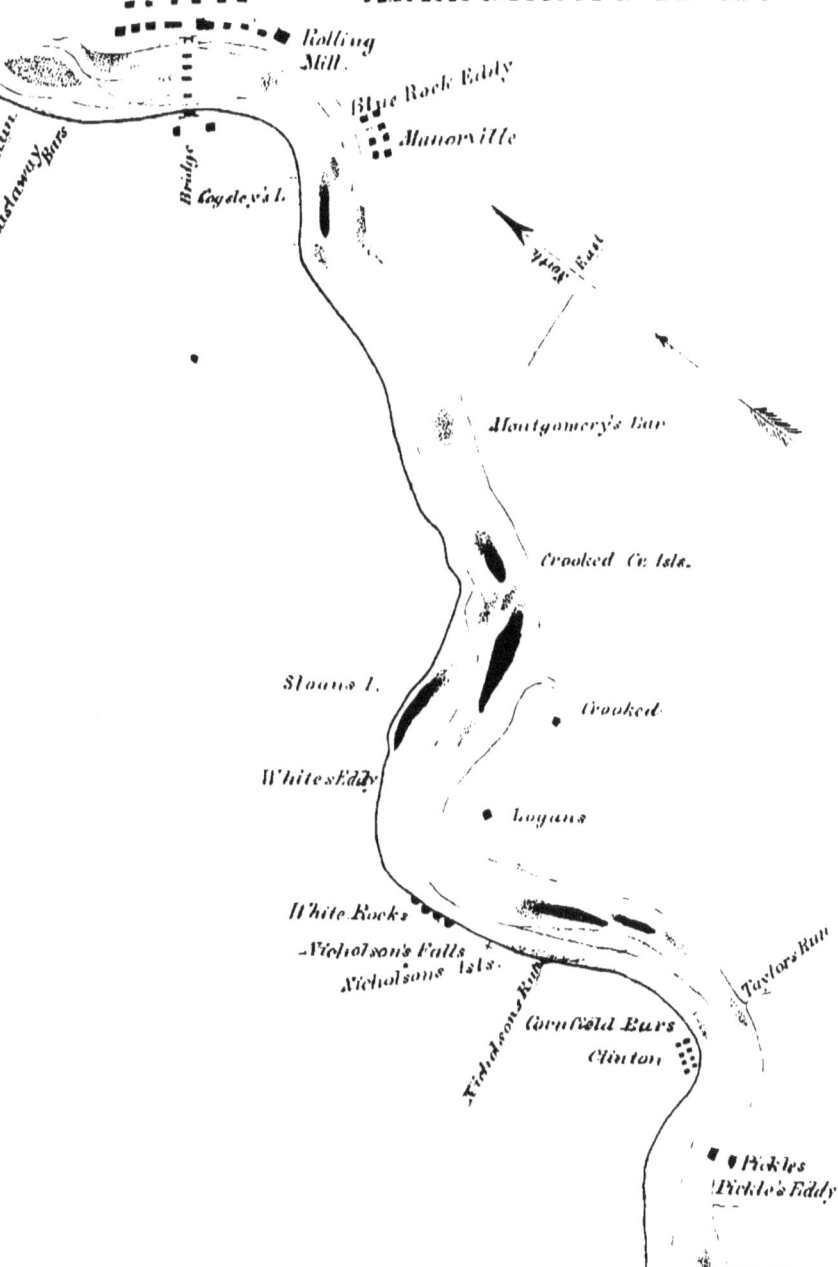

ALLEGHENY No 13.

LLEGHNY No. 14.

Mad Dog Riffle
Sugar Loar Rock
Salt Works
Murphy's I.
Carding Mill Run
Mad Dog I.
Aqueduct
Kiskiminitas Cr.
Rail Road Bridge
Freeport I.
Freeport
Locust Run
Buffalo Cr.
Harbison Run
McCains
East
Karn's I.
Walts Run
Jacks I.
Soda Works
Judge Brackenridge
Chartier Cr.

ALLEGHENY No. 15.

Bull Creek I.
Salt Works
Lower Salt Works
Bull Cr. Bars
Cole's Run
Puckerty I.
Pocketas
Coal
Logans
Logan's Eddy
Tawny Hill Run
North East
White Rock
Panther Run
14 Mile I.
Black's Run
Falling Spring
12 Mile I.
Deer Cr.
Plum Cr.
Mechanicsburg
Plum Cr. Bars
Hulings Eddy
Hulings
Powers Run
Fairview
Quigley's Run
9 Mile Isl.

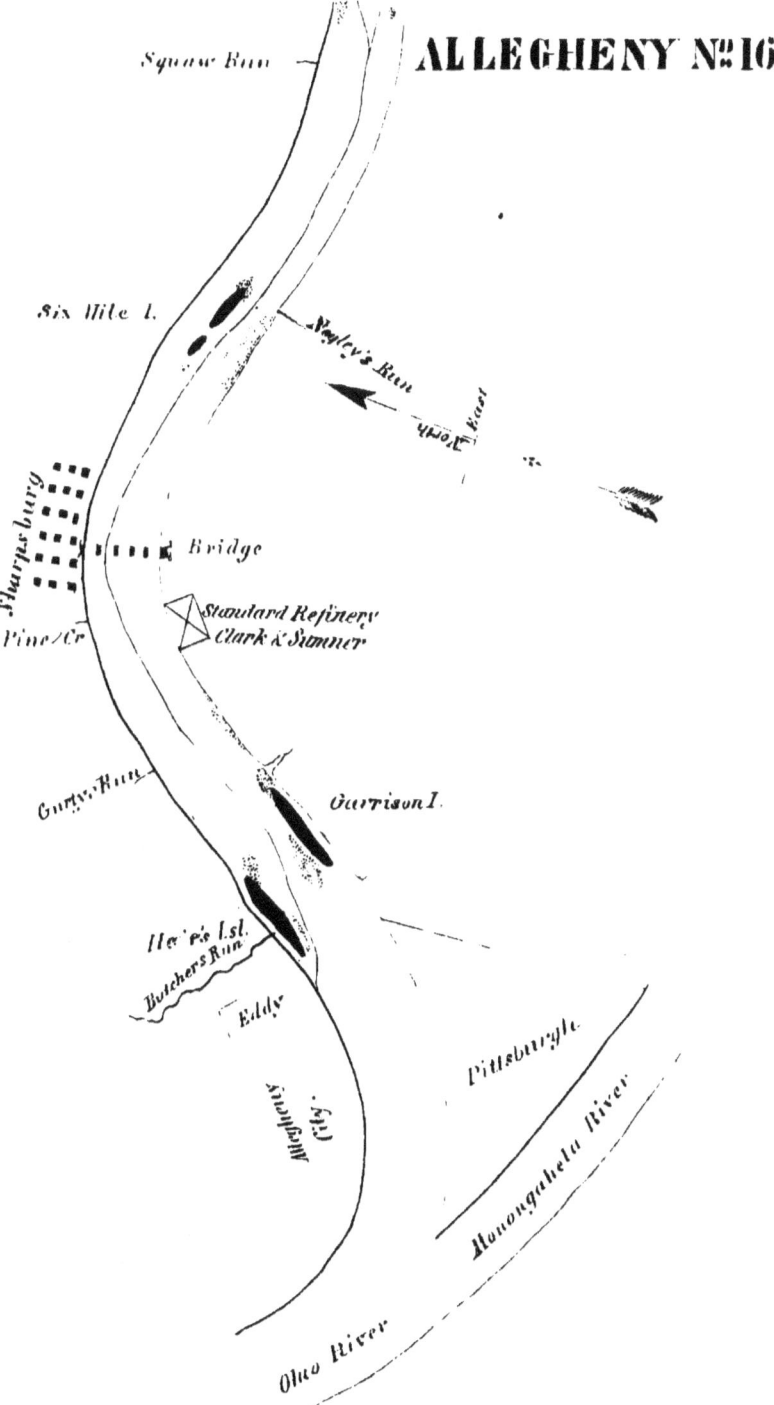

DIRECTIONS FOR MAP NO. 16.

Six-Mile Island.

[Warren, 195¼ miles—Pittsburg, 7¾ miles.]

CHANNEL to the left. Enter the chute to the right of the middle; afterward incline to the right, and pass near the foot of the island, to avoid a large bar on the left, below the mouth of Negley's Run.

Sharpsburg Bridge.

[Warren, 197½ miles—Pittsburg, 5¾ miles.]

Channel second space from the right shore.

Hare's Island, on the Right.

[Warren, 200¼ miles—Pittsburg, 2¾ miles.]

Garrison Island on the left; channel about midway. After passing the head bar of Hare's Island, keep to the right, and run close to the foot, in order to either land in the eddy below the island, or to pass the bridges.

In passing the bridges, it is the usual custom to take the first or second spans from the right shore. After passing the lowermost bridge, incline outward, so as to be to the left of the middle when opposite the

Point at Pittsburg.

[Warren, 203 miles.]

PITTSBURG.

ITS LOCATION.

PITTSBURG, the second city of Pennsylvania, is situated at the head of the Ohio River, formed by the junction of the Monongahela and Allegheny Rivers—the Allegheny running from the north-east, and the Monongahela from the south-west, meeting at an angle of about thirty-three degrees.

ITS EARLY HISTORY.

Pittsburg has early associations in its history. Not only the early settlers of America, but even France and England, regarded the site a very important one to sustain the occupation of this department of a new country. As early as 1784, mention has been made by historians that the ground between Fort Pitt and the Allegheny River was disposed of by the original proprietaries. The sale embraced what was supposed to be about three acres, which was soon laid out in town-lots. This disposition of the land, no doubt, was influenced by the importance of the situation, already ratified by military operations, which have become historic. The visits of distinguished personages, for civic and military purposes, at this period, gave character and *eclat* to the embryo city. Very soon merchandising and manufacturing, with capital correspondent with the size and wants of the place, were established. Then came the more powerful civilizers—the school-house, the press, and the church. Connection with the East was transitory and unreliable—sometimes by pack-mule transportation, at other times by an occasional traveler, who venturously climbed the Allegheny, to take a glimpse at the opening of the Mississippi Valley.

Five hundred inhabitants, in 1788, only four years subsequent to its original laying out, with very few, if any, of the comforts of civilized life, congregated together, remote from social ties, prognosticated already the future greatness and prosperity which the present generation have seen verified.

The small manufacturing establishment of that early day was but the parent of the present large establishments, which, in their proportions and character of manufacture, so justly entitles them to rank with a Birmingham or a Sheffield of another country.

ITS WATER AND RAILROAD FACILITIES.

Water and railroad communications have created avenues for trade and intercourse in every direction throughout the country. Situated at the head of the Ohio River, it enjoys direct communication with the Mississippi Valley, embracing a region of thousands of miles, with its cities, its towns, and its agricultural districts, necessarily requiring the manufactures of this industrial city.

The Monongahela, as it courses its way through fertile districts, embraces enterprise and industry, with which Pittsburg can exchange with reciprocal benefits.

Again, the Allegheny River, with its rich agricultural valley, presents another communication by which the products of the farm can be exchanged for the products of the mine and the shop; while the oil region, with its immense wealth, situated upon this stream and its tributaries, can find a market or an outlet to all parts of the United States, through the manifold water and river communications which throw their branches throughout so much of that territory; even oceanic transportation may be obtained.

While river transportation is directed toward so many districts, the interior of the great Western States is brought near by railroads, and the Eastern cities visited with speedy and safe carriage by similar conveyances, affording communication with the seaboard. Large quantities of oil are now daily carried over the railroad to the East. The Allegheny Valley, with other small contiguous valleys, seem to be the actual oil region. This is predicated upon the fact that its products furnish the demand. The oil, to a great extent, finds its way to Pittsburg by the Allegheny River, which, when in a very low stage, is subserved by railroad transportation.

POPULATION AND BUSINESS.

The resident population of Pittsburg is about eighty thousand. Birmingham and Allegheny cities are connected with Pittsburg by bridges, both of which are flourishing towns, and have many manufactories of iron, glass, and wares of other kinds, producing a large quantity yearly.

There are in Pittsburg and its vicinity eighteen rolling-mills, with a large amount of capital invested, and employing some thousands of hands, besides

many very large foundries, together with **many
smaller ones, employing a very large number of
hands and capital.**

The cotton factories, the glass, the white-**lead, the
manufactories of axes, hatchets, steel-springs, anvils,**
axles, vises, saws of every kind; shovels, **spades,**
forks, hoes, and every useful and ornamental article
of iron which is now used, employ immense capital,
which deservedly entitles this city to the rank of **the**
greatest manufacturing power of America.

The oil which daily descends the Allegheny, if in
its crude state, finds a market. Large and capacious
refineries are in operation, and add a boundless store
of wealth and employ to the city.

The levee of the Allegheny, with its boats unlading, with its tiers of barrels arrayed with business-like precision, and the large storehouses and refineries fronting the river, present a scene of activity which, in its wealth, enhances the already great prosperity of the Iron City.

BUSINESS CHARACTER.

The industrial energy and spirited character of its people often astounds a visitor. Here everybody works, not by proxy, but *in propria persona.* The head of a wealthy manufacturing or mercantile establishment may be often seen, in business hours, minus his coat and vest, now heaving a box, or a heavy piece of iron, with his porter's help; again, rolling a box on a dray, with the muscular power of the strong drayman himself.

The timidity lest such work may be unfashionable, as voted in many mercantile circles of even less proportion than the Smoky City, is here changed to a courage noble, manly, and unselfish. This *fashion* has made the city. It has made wealth. It has supplied the country with millions of necessaries, for which, in the absence of industrial care and labor, humanity would have been less benefited.

The whole number of refineries in the city and vicinity is fifty-eight, with a capacity per week of nine thousand barrels.

Value of real estate, building, and machinery, $2,534,000; value of refined oils, $8,599,223; wages paid per annum, $350,000.

PETROLEUM.

The sources of petroleum and mineral pitch in Asia and other countries are confined to the rocks of new, secondary, and tertiary ages, which in the Alleghany regions furnish such abundant supplies of petroleum.

Bitumen, in the paleozoic basin of North America, either in liquid or solid, depends upon oxygen, since, by oxydation, the petroleum becomes slowly changed into naphtha, which, at ordinary temperature, is solid.

In the calcareous sand-rocks in New York, a black substance, which has been called anthracite, has been found, in which the bitumen appears to have lost its fusibility, and even its solubility.

A similar material occurs in the Quebec group in Canada, which evidently showed that at some ancient period it was semi-fluid.

It is very easy to distinguish between lignite and bituminous rocks. Some are disposed to regard the former as the source of bitumen. Later examinations, however, show that bitumen has been generated under different conditions from those which have transformed organic matter into coal.

We find in the Utica and Hamilton formations highly inflammable pyro-schists, which contain no soluble bitumen; but yet the Trenton limestone is impregnated with petroleum, and gave rise to many petroleum springs.

It has now been safely concluded that petroleum has been generated by the transformation of organic matter in underlying strata.

Some theories have been advanced that coal is the product of mineral substances, and that petroleum was distilled from coal. If this should be true, how could carbon acquire such a condition of purity if coal was the first product? There are some who trace the change of wood to peat, brown coal, and bituminous coal, through analysis.

T. Sterry Hunt, M.A., F.R.S., when upon a geo-scientific research, and now refutes this theory in his logical survey of Canada, gave the country a thorough "Notes on the History of Petroleum, or Rock Oil."

LIST OF OIL COMPANIES.

PITTSBURG.

Names of Companies.	Capital.	Number of Shares.	Par Value
Acme	$200,000	10,000	...
Allegheny and Pittsburg			...
Ardesco			...
Banner	130,000	65,000	...
Blood Farm and Ohio Valley			...
Cherry Run Central	250,000	250,000	...
Citizens'			...
Columbia	2,500,000	50,000	...
Culbertson Run	150,000	75,000	...
Fayette			...
Federal	480,000	60,000	...
Fleming & Blood	100,000	20,000	...
Germania	200,000	40,000	...
Good Intent			...
Horse Neck	100,000	50,000	...
Iron City	150,000	30,000	...
Linden	200,000	40,000	...
Luseesco			...
McAboy Cherry Run			...
Merchants'	50,000	100,000	...
National			...
Newal's Run			...
Nonpareil			...
North American			...
Ohio Valley	200,000	40,000	...
Pittsburg and Great Western	100,000		...
Pittsburg and West Virginia			...
Ross	100,000	100,000	...
Smoky City			...
Stella			...
Story	25,000	50,000	...
Venango Central	75,000	75,000	...
West Virginia	20,000	20,000	...
West Virginia Basin	20,000	20,000	...
Whitely Creek			...

NEW YORK.

Names of Companies.	Capital.	Number of Shares.	Par Value
American	$500,000		$100
Atlantic	300,000	30,000	10
Bergen	2,000,000	200,000	10
Blood Farm	5,000,000	50,000	100
Buchanan Farm	4,000,000	400,000	10
Buchanan Oil and Rectifying	200,000	40,000	5

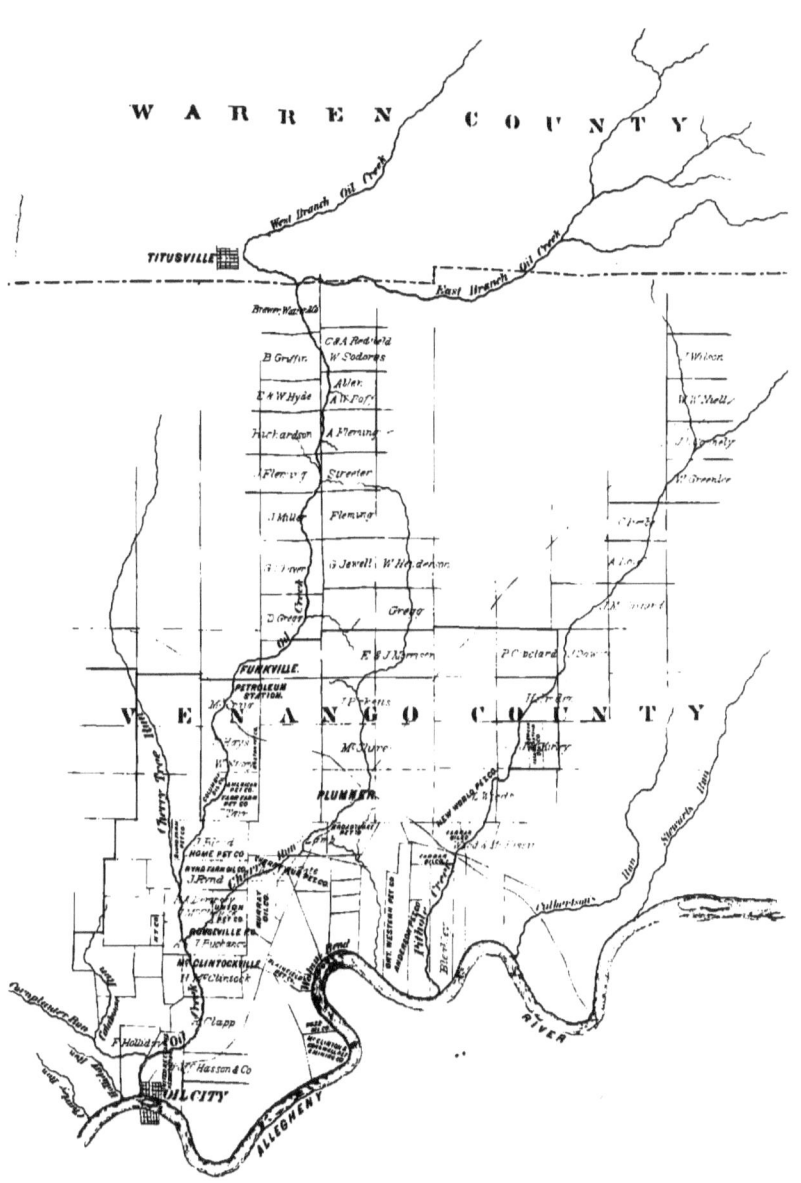

LIST OF OIL COMPANIES.

NEW YORK.
(CONTINUED.)

Names of Companies.	Capital.	Number of Shares.	Par Value
Central	$5,000,000	50,000	...
Clinton	500,000	50,000	$10
Commercial	5,000,000	50,000	100
Consolidated (of N. Y.)	1,000,000	100,000	10
Enterprise	100,000	10,000	10
Flowing Well	50,000	1,000	50
Great Western	....	....	...
Highgate	....	....	...
Home	5,000,000	50,000	100
Hope	....	....	...
Island	500,000	50,000	10
Kanawha	500,000	50,000	10
Knickerbocker	....	....	...
Manhattan	....	....	...
Maple Shade (of N. Y.)	500,000	50,000	10
Marietta	....	....	...
Miller Farm	300,000	60,000	5
McKinley	250,000	25,000	10
New York and West Virginia	500,000	50,000	10
New York and Pennsylvania	1,000,000	100,000	10
Noble Well	500,000	20,000	25
Oil Creek	1,000,000	50,000	20
Revenue	250,000	5,000	50
Rynd Farm	2,000,000	200,000	10
Story & McClintock	....	....	...
Success	100,000	10,000	10
Union	125,000	12,500	10
United States	500,000	50,000	10
Vesta Petroleum and Refining	500,000	5,000	100

PHILADELPHIA.

Names of Companies.	Capital.	Number of Shares.	Par Value
Ætna	$500,000	50,000	$10
Adamantine Oil Co	500,000	50,000	10
Alcorn Oil Co	500,000	50,000	10
Allegheny River Oil Co	75,000	75,000	1
Allegheny and Pittsburg	500,000	50,000	10
Allegheny and Tidionte Oil	....	....	...
Allegheny and Walnut Bend	250,000	50,000	5
American Oil Co	500,000	50,000	10
Beacon Oil	500,000	50,000	10
Big Tank Oil Co	500,000	100,000	5
Bleakley Oil Co	500,000	50,000	10

PHILADELPHIA.

(CONTINUED.)

Names of Companies.	Capital.	Number of Shares.	Par Value
Briggs	$500,000	50,000	$10
Bruner Oil and Mining	500,000	100,000	5
Bull Creek Oil Co	1,000,000	100,000	10
Cherry Run Oil Co.	200,000	20,000	10
Columbia Oil Co., Pittsburg	2,500,000	250,000	10
Commercial Oil	500,000	50,000	10
Consolidated (of Philadelphia)	600,000	60,000	10
Continental	500,000	50,000	10
Cornplanter	1,000,000	100,000	10
Cow Creek and Stilwell Run	500,000	100,000	5
Curtin	500,000	50,000	10
Delzell Oil	2,000,000	200,000	10
Dark Hollow	500,000	50,000	10
Decatur	400,000	40,000	10
Densmore	500,000	50,000	10
Diamond	500,000	50,000	10
Eagle Rock Oil Co	500,000	50,000	10
Egbert Oil	500,000	50,000	10
Empire and Oil City	500,000	50,000	10
Excelsior Oil	500,000	100,000	5
Farel	500,000	50,000	10
Franklin Lubricating Oil Co	500,000	25,000	20
French Creek Lubricating	450,000	30,000	15
Girard	100,000	10,000	10
Globe	300,000	30,000	10
Great Western Oil Co., N. Y.	500,000	50,000	10
Green Hill	500,000	50,000	10
Hibbard Oil Co	500,000	100,000	5
Hogo Island	250,000	25,000	10
Holland Petroleum Co.	500,000	50,000	10
Hoover	500,000	50,000	10
Hope Farm	....	....	...
Horse Creek	500,000	50,000	10
Hosmer Oil Co	500,000	50,000	10
Howe's Eddy	500,000	50,000	10
Hughes River	300,000	30,000	10
Irwin	500,000	50,000	10
Junction	500,000	50,000	10
Keystone	500,000	50,000	10
Lamberton Oil and Manufacturing Co.	125,000	25,000	5
Lancaster	500,000	50,000	10
Maple Shade (of Philadelphia)	500,000	50,000	10
McClintockville	2,000,000	200,000	10
McCormick & McKissock	500,000	50,000	10
McElheny	1,000,000	100,000	10
McGuire	500,000	50,000	10

LIST OF OIL COMPANIES. 67

PHILADELPHIA.

(CONTINUED.)

Names of Companies.	Capital.	Number of Shares.	Par Value
Metropolitan	....	....	$10
Middle Walnut Oil Co	$500,000	50,000	10
Mineral	500,000	50,000	10
Noble & Delamater	500,000	50,000	10
Oil Creek Petroleum	500,000	50,000	10
Oil Valley Petroleum (of Penn.)	375,000	37,500	10
Olmstead	500,000	50,000	10
Ormsby Petroleum Co	500,000	50,000	10
Organic	500,000	50,000	10
Parker Petroleum Co	....	....	...
Pearson Petroleum Co	500,000	100,000	5
Pennsylvania Central Oil Co	200,000	20,000	10
Pennsylvania Oil Creek Petroleum	500,000	50,000	10
Perry	500,000	50,000	10
Petroleum Center	500,000	50,000	10
Philadelphia Oil	250,000	10,000	25
Philadelphia and Oil Creek	500,000	50,000	10
Philadelphia and Tidioute Island	500,000	50,000	10
Phillips	200,000	40,000	5
Pit Hole	500,000	50,000	10
Pittsburg and Oil Creek Petrona	500,000	50,000	10
Pope Farm Oil	500,000	50,000	10
Quaker City	100,000	10,000	10
Reliance	500,000	50,000	10
Revenue Oil Co	250,000	50,000	25
River Oil	500,000	50,000	10
Roberts' Oil and Mining	500,000	50,000	10
Rock	500,000	50,000	10
Rockland	500,000	50,000	10
Schuylkill and Oil Creek	500,000	50,000	10
Seneca	500,000	25,000	20
Story Farm	1,000,000	200,000	5
Story Center Oil Co	500,000	50,000	10
Sugar Creek	500,000	50,000	10
Sutle Lubricating Oil Co	500,000	50,000	10
Tarr Farm Oil Co	250,000	50,000	5
Tidioute and Allegheny	250,000	50,000	5
Tipton	500,000	50,000	10
Union	500,000	100,000	5
Upper Economy Petroleum Co	500,000	50,000	10
Van Buren Oil Co	500,000	25,000	20
Vance Stewart	300,000	60,000	5
Venango	500,000	25,000	20
West Virginia	200,000	20,000	10
Washington Oil Co	500,000	25,000	20
Watson Petroleum Co	500,000	50,000	10

DUFF'S COLLEGE,
Iron Block, No. 37 Fifth Street.

DUFF'S ORIGINAL PLAN OF
BUSINESS EDUCATION,
Taught for the last twenty-five years from his systems of
MERCANTILE, BANK, RAILROAD & STEAMBOAT BOOK-KEEPING,
With First Premium **BUSINESS** and **ORNAMENTAL PENMANSHIP.**
The only Institution of the kind in the United States conducted
by a PRACTICAL MERCHANT.

HARPER'S ENLARGED EDITION OF
DUFF'S MERCANTILE BOOK-KEEPING.
Sold by Booksellers generally. Price $1.87. Postage 20 Cents.
Sanctioned by Special Committees of the American Institute and New York Chamber of Commerce, as the most thorough system of accounts published.

AWARDED FOUR SILVER MEDALS,

Which, with the following testimonials indicate the character of this work:

"No other work upon Book-keeping explains the subject with so much clearness and simplicity."—F. W. EDMONDS, *Cashier Mechanics' Bank, Wall st., N. Y.*

"It gives a clear insight into all departments of this science."—A. S. FRASER, *Cashier of Seventh Ward Bank, N. Y.*

"As an extensive ship owner, American and European merchant, bank director, etc., he has born the reputation of the highest order of business talents."—JNO. W. BURNHAM, *Merchant, No. 8 South street, N. Y.*

"Mr. Duff is a man of rare qualifications for business."—JOHN M. D. TAYLOR, *Merchant, Union street, New Orleans.*

"Mr. Duff is a merchant of the first respectability."—J. LANDIS, *Merchant, New Orleans.*

"I graduated in Duff's College in half the time I expected. His admirable system includes nothing superfluous, nor leaves out anything essential."—J. R. COMPTON, *Cashier Niagara Bank, Lockport, N. Y.*

"It contains much matter important to the merchant."—C. O. HALSTEAD *President Manhattan Bank, N. Y.*

"The most complete work of the kind I have ever seen."—JAS. B. MURRAY, *President Exchange Bank, Pittsburgh.*

"The most clear and comprehensive that I have met with."—JOHN SNYDER, *Cashier Bank Pittsburgh.*

"You have put your own long experience as a merchant to good use in this work."—RICHARD IRVIN, *Merchant, No. 98 Front street, N. Y.*

"The favorable opinions already expressed by gentlemen of competent authority, are well deserved, and in this case very properly bestowed."
CHARLES M. LEUP, } Special Committee of the
LEOPOLD BIERWORTH, } Chamber of Commerce, N. Y.
ROBERT KELLY,

Extract from the Minutes.
PROSPER M. WETMORE, Secretary.

"Your Committee unanimously concur in the opinion of the utility of the improved method of Mr. Duff."—GURDON J. LEEDS, *Recording Secretary of the American Institute, New York.*

A NEW ENLARGED EDITION OF
DUFF'S STEAMBOAT BOOK-KEEPING,
Just published by the Author. Sold by Booksellers generally. Price $2.00
Pronounced by a competent Pittsburgh Accountant, "A PERFECT SYSTEM FOR SUCH BOOKS AND ACCOUNTS."

"The form of the Steamers Protest is worth more than the price of the Book. I never leave port without a copy of it aboard."—A. C. McCALLAM, *Captain steamer Arcola.*

"There is nothing else on the subject published, of any value to the steamer's accountant."—J. F. J. ALLISON, *late First Clerk of steamer Fortune.*

☞ Our Bank and Railroad Books are used in Manuscripts only. ☜ For full particulars, send for our elegant new Circular of 75 pages, octavo, with samples of our Penman's writing, enclosing 25 cents to

P. DUFF & SON.
Principals.

PITTSBURGH FEMALE COLLEGE,

REV. I. C. PERSHING, D. D., President.

Best sustained College in the State. ATTENDANCE PRESENT YEAR, UPWARDS OF FOUR HUNDRED. Large and beautiful brick buildings, with all the modern improvements. Twenty able and accomplished Teachers

FRENCH AND GERMAN
Are taught by Native Teachers.

THE DRAWING AND PAINTING DEPARTMENT
Is very large, and is under the care of an accomplished Artist from New York, who has been connected with the College for SEVEN YEARS.

THE MUSICAL DEPARTMENT,

Numbering, the past year, upwards of TWO HUNDRED, is under the care of PROF. HENRY ROHBOCK, a German gentleman of rare musical talents, who has devoted his entire life-time to his profession, and who has had UPWARDS OF TWENTY-FIVE YEARS EXPERIENCE IN TEACHING. Four additional teachers are employed in the same Department. In this, as in all the other Departments, it is confidently believed that the College offers advantages unsurpassed by any similar institution in the land.

THE BOARDING DEPARTMENT

Is under the special care of the President. FIFTY DOLLARS per term—of fourteen weeks—pays all expenses in this Department, except fuel and washing.

FALL TERM commences Tuesday August 29th, 1865.
WINTER TERM, Tuesday December 5th, 1865.
SPRING TERM Thursday March 23d, 1866.
Send to President Pershing for a Catalogue.

M. SIMPSON.
Pres. Trustees.

PITTSBURGH, May, 1865.

PITTSBURGH
GENERAL INSURANCE AGENCY,

No. 64 FOURTH STREET.

COMPANIES REPRESENTED:

ÆTNA,	Hartford,	Assets,............................$4,000,000.
CHARTER OAK,	"	"	375,000.
HARMONY,	New York,	"	500,000.
THAMES,	Norwich,	"	250,000.

These old and popular Companies, having aggregate cash assets exceeding

FIVE MILLIONS OF DOLLARS,

Continue to accept risks on all descriptions of property, including

Refineries, Petroleum and its Products,

Either in Warehouses, Tanks, Yard, or on Wharf, *and in transit*,

ON REASONABLE TERMS.

Policies issued without delay, and all business attended to with Fidelity and Dispatch, on application to

A. A. CARRIER & BRO.,
Agents and Attorneys.

B. L. H. DABBS,
DEALER IN
Photographic Materials
OF EVERY DESCRIPTION,
No. 52 St. Clair Street, opposite St. Clair Hotel,
PITTSBURG, PA.

All Chemicals, Paper and Cameras thoroughly tested at my Photographic Galleries, Nos. 46, 48 and 50 ST. CLAIR ST.

Always on hand an extensive and complete assortment of

Cameras, Apparatus, Cases, Chemicals, Glass, Plates, Matting, Passe-Partouts, Preservers, Photographic Paper, Back-Grounds, Sky-Lights, Chairs, &c., &c.

Orders by Mail or Express attended to with Promptness and Dispatch.

PHOTOGRAPHIC GALLERY,
46, 48 and 50 ST. CLAIR ST.
PITTSBURG, PA.

WATER COLOR MINIATURES, LIFE-SIZE PHOTOGRAPHS IN CRAYON, OIL, INDIA INK, etc. PHOTOGPAPHS WITH LANDSCAPE AND FANCY BACKGROUNDS.
IVORYTYPES.

PITTSBURGH,   MAY, 1865.

GENERAL FOREIGN AGENCY
AT THE
ADAMS' EXPRESS OFFICE.

Passage to and from Europe in Sailing Vessels.
Drafts on all parts of Europe.

                           WM. BINGHAM, Jr.,
                               ADAMS EXPRESS OFFICE, 54 Fifth Street,
                                             PITTSBURGH, PA.

WHITMORE, WOLF, DUFF & CO.,

SIGN OF THE ANVIL.

IMPORTERS AND DEALERS IN

Hardware and Cutlery,

Three doors above St. Charles Hotel,
50 WOOD STREET,

Pittsburgh.

M. WHITMORE,   C. H. WOLF,
GEO. J. DUFF,   THOS. B. LANE.

B. WOLF, JR.
Importer and Dealer in
HARDWARE,

Consisting in part of

BLACKSMITHS', CARPENTERS', SADDLERS' AND COOPERS' TOOLS; LOCKS, LATCHES, MILL AND CROSS-CUT SAWS, AXES AND STEEL, RIFLE BARRELS, GUN TRIMMINGS, SADDLERY, HARDWARE, PLATFORM SCALES, CUTLERY, COFFIN TRIMMINGS, LACE LEATHER.

Corner of Liberty and St. Clair Streets,
PITTSBURGH, PA.

www.ingramcontent.com/pod-product-compliance
Lightning Source LLC
Chambersburg PA
CBHW030405170426
43202CB00010B/1497